T0327531

EVOLUTIONARY TOPOLOGY OPTIMIZATION OF CONTINUUM STRUCTURES

EVOLUTIONARY TOPOLOGY OPTIMIZATION OF CONTINUUM STRUCTURES

METHODS AND APPLICATIONS

X. Huang

School of Civil, Environmental and Chemical Engineering,
RMIT University, Australia

Y.M. Xie

School of Civil, Environmental and Chemical Engineering,
RMIT University, Australia

A John Wiley and Sons, Ltd., Publication

This edition first published 2010
© 2010, John Wiley & Sons, Ltd

Registered office
John Wiley & Sons Ltd, The Atrium, Southern Gate, Chichester, West Sussex, PO19 8SQ, United Kingdom

For details of our global editorial offices, for customer services and for information about how to apply for permission to reuse the copyright material in this book please see our website at www.wiley.com.

Library of Congress Cataloging-in-Publication Data

Huang, X. (Xiaodong), 1972–
   Evolutionary topology optimization of continuum structures : methods and applications / by X. Huang, Y.M. Xie.
      p. cm.
   Includes bibliographical references and index.
   ISBN 978-0-470-74653-0 (cloth)
1. Structural optimization.   2. Topology.   I. Xie, Y. M.   II. Title.
   TA658.8.H83 2010
   624.1′7713–dc22

                                                                                2009049233

A catalogue record for this book is available from the British Library.

ISBN: 978-0-470-74653-0 (Hbk)

Typeset in 10/12pt Times by Aptara Inc., New Delhi, India

Contents

Preface

Since the late 1980s, enormous progress has been made in the theory, methods and applications of topology optimization. Among various numerical methods for topology optimization, Evolutionary Structural Optimization (ESO) and Bi-directional Evolutionary Structural Optimization (BESO) have been extensively investigated by many researchers around the world. The first book on ESO was published by Y.M. Xie and G.P. Steven in 1997. Since then the field has experienced rapid developments with a variety of new algorithms and a growing number of applications.

There are many different versions of ESO/BESO algorithms proposed by several dozens of researchers in the past two decades. However, some of the algorithms appeared in the literature are unreliable and inefficient. The primary purpose of this book is to provide a comprehensive and systematic discussion on the latest techniques and proper procedures for ESO/BESO, particularly BESO, for the topology optimization of continuum structures.

The BESO method is presented here for a wide range of structural design problems including stiffness and frequency optimization, nonlinear material and large deformation, energy absorption, multiple materials, multiple constraints, periodical structures, and so on. Numerous examples are provided to demonstrate the efficacy of the techniques and the applicability to real structures.

This book is written for researchers and engineers, both academic and practising, with an interest in structural optimization. Their disciplines include civil, mechanical, aerospace and biomedical engineering. The material contained in the book will also be useful to architects seeking to create structurally efficient and aesthetically innovative buildings and bridges. For further information, please visit http://www.wiley.com/go/huang

The bulk of the material presented in this book is the result of the authors and their coworkers, mainly since 2004. The authors would like to acknowledge their considerable debt to all those who have contributed to the work, particularly T. Black, J. Burry, M.C. Burry, D.N. Chu, P. Felicetti, Y.C. Koay, K. Ghabraie, Q. Li, Q.Q. Liang, G. Lu, A. Maher, O.M. Querin, G.P. Steven, J.W. Tang, X.Y. Yang and Z.H. Zuo. The authors are grateful to H. Ohmori for providing digital images for two interesting examples included in Chapter 9. Thanks are also due to Y.X. Yao and S.R. Guillow who read the manuscript at various stages and made many valuable suggestions for improvement.

The authors wish to express their special gratitude to G.I.N. Rozvany. Over the last decade, through a series of publications, he has provided the most insightful observations and suggestions about ESO/BESO, which have inspired the authors to make significant modifications and enhancements to their algorithms in recent years. Without his deep insight, much of the material presented in this book would not have been possible.

Xiaodong Huang and Mike Xie
Melbourne, August 2009

1

Introduction

1.1 Structural Optimization

Structural optimization seeks to achieve the best performance for a structure while satisfying various constraints such as a given amount of material. Optimal structural design is becoming increasingly important due to the limited material resources, environmental impact and technological competition, all of which demand lightweight, low-cost and high-performance structures.

Over the last three decades the availability of high-speed computers and the rapid improvements in algorithms used for design optimization have transformed the topic of structural optimization from the previous narrowness of mostly academic interest to the current stage where a growing number of engineers and architects start to experiment with and benefit from the optimization techniques. There have been more and more research and development activities directed towards making the structural optimization algorithms and software packages available to the end-users in an easy, reliable, efficient and inexpensive form.

The types of structural optimization may be classified into three categories, i.e. size, shape and topology optimization. Size optimization is to find the optimal design by changing the size variables such as the cross-sectional dimensions of trusses and frames, or the thicknesses of plates. This is the easiest and earliest approach to improving structural performance. Shape optimization is mainly performed on continuum structures by modifying the predetermined boundaries to achieve the optimal designs. Topology optimization for discrete structures, such as trusses and frames, is to search for the optimal spatial order and connectivity of the bars. Topology optimization of continuum structures is to find the optimal designs by determining the best locations and geometries of cavities in the design domains. It is worth pointing out that all the topology optimization methods considered in this book can be readily used to perform shape optimization by simply restricting the structural modification to the existing boundaries.

1.2 Topology Optimization of Continuum Structures

Compared with other types of structural optimization, topology optimization of continuum structures is by far the most challenging technically and at the same time the most rewarding

Evolutionary Topology Optimization of Continuum Structures: Methods and Applications   Xiaodong Huang and Mike Xie
© 2010 John Wiley & Sons, Ltd

economically. Rather than limiting the changes to the sizes of structural components, topology optimization provides much more freedom and allows the designer to create totally novel and highly efficient conceptual designs for continuum structures. Not only can the topology optimization techniques be applied to large-scale structures such as bridges and buildings (e.g. Cui *et al.* 2003; Ohmori *et al.* 2005), they may also be used for designing materials at micro- and nano-levels (e.g. Sigmund 1995; Torquato *et al.* 2002; Zhou and Li 2008).

Starting with the landmark paper of Bendsøe and Kikuchi (1988), numerical methods for topology optimization of continuum structures have been investigated extensively. Most of these methods are based on finite element analysis (FEA) where the design domain is discretized into a fine mesh of elements. In such a setting, the optimization procedure is to find the topology of a structure by determining for every point in the design domain if there should be material (solid element) or not (void element). In recent years, topology optimization has become an extremely active area of research and development. Hundreds of publications have emerged, including a number of books, e.g. Bendsøe (1995); Xie and Steven (1997); Hassani and Hinton (1999); Bendsøe and Sigmund (2003). As a result, several numerical methods of topology optimization have reached the stage of practical applications including the SIMP method and the ESO method. The term 'SIMP' stands for Solid Isotropic Material with Penalization for intermediate densities. The original idea of the SIMP method was proposed by Bendsøe (1989). 'ESO' stands for Evolutionary Structural Optimization, which is a design method based on the simple concept of gradually removing inefficient material from a structure.

1.3   ESO/BESO and the Layout of the Book

The literature on ESO is most extensive, with well over a hundred publications (starting with Xie and Steven 1992). The first book on ESO (Xie and Steven 1997) summarized the early developments of the technique. Since then significant progress has been made in improving the algorithms of ESO and Bi-directional Evolutionary Structural Optimization (BESO). At the same time, there is a great deal of confusion among researchers in the structural optimization community with regard to the efficacy of ESO/BESO because some of the early versions of ESO/BESO algorithms did not adequately address many important numerical problems in topology optimization. This book provides a comprehensive and systematic discussion on the latest techniques and proper procedures for ESO/BESO, particularly BESO, for topology optimization of continuum structures.

Chapter 2 briefly describes of the original ESO method based on the elemental stress level, followed by a discussion on the ESO technique for stiffness optimization.

Chapter 3 introduces a new BESO algorithm for stiffness optimization which addresses many issues related to topology optimization of continuum structures such as mesh-dependency, checkerboard pattern, and convergence of solutions.

Chapter 4 develops a soft-kill BESO method utilizing a material interpolation scheme with penalization for intermediate densities. To some extent, this is similar to the SIMP method. However, one important difference is that the soft-kill BESO uses discrete design variables while SIMP allows for continuous material densities. At the end of the chapter, a computer code of BESO written in MATLAB® is provided.

Chapter 5 presents a comparison between the ESO/BESO methods and the SIMP method. Also included in the comparison is a more sophisticated algorithm of the SIMP approach called the continuation method.

Chapter 6 extends the BESO method to a range of topology optimization problems including minimizing weight with a displacement or compliance constraint, maximizing stiffness with an additional displacement constraint, stiffness optimization with multiple materials and multiple load cases, optimal design of periodical structures, optimization for design-dependent loading, maximizing natural frequencies, as well as a BESO algorithm based elemental stress level.

Chapter 7 discusses topology optimization of nonlinear continuum structures. Both material nonlinearity and large deformation are considered. The loading on the structure can be either prescribed forces or prescribed displacements at specific locations.

Chapter 8 expands the technique of topology optimization of nonlinear structures by developing a BESO algorithm for the optimal design of energy-absorbing structures.

Chapter 9 conducts case studies on several practical applications of the BESO method to demonstrate the potential benefit of employing such a technique.

Chapter 10 introduces an easy-to-use computer program called BESO2D which is can be used by the reader as well as students to perform stiffness optimization of two-dimensional structures.

Various other BESO software packages based on the work described in this book, including a source code written in MATLAB, can be downloaded from the website www.isg.rmit.edu.au, or obtained from the authors by emailing huang.xiaodong@rmit.edu.au or mike.xie@rmit.edu.au.

References

Bendsøe, M.P. (1989). Optimal shape design as a material distribution problem. *Struct. Optim.* **1**: 193–202.

Bendsøe, M.P. (1995). *Optimization of Structural Topology, Shape and Material.* Berlin: Springer.

Bendsøe, M.P. and Kikuchi, N. (1988). Generating optimal topologies in structural design using a homogenization method. *Comput. Meth. Appl. Mech. Engng.* **71**: 197–224.

Bendsøe, M.P. and Sigmund, O. (2003). *Topology Optimization: Theory, Method and Application*, Berlin: Springer.

Cui, C., Ohmori, H. and Sasaki, M. (2003). Computational morphogenesis of 3D structures by extended ESO method. *J. Inter. Assoc. Shell Spatial Struct.* **44**(1): 51–61.

Hassani, B. and Hinton, E. (1999). *Homogenization and Structural Topology Optimization*, Berlin: Springer.

Ohmori, H., Futai, H., Iijima, T., Muto, A. and Hasegawa, H. (2005). Application of computational morphogenesis to structural design. In Proceedings of Frontiers of Computational Sciences Symposium, Nagoya, Japan, 11–13 October, 2005, pp. 45–52.

Sigmund, O. (1995). Tailoring materials with prescribed elastic properties. *Mech. Mater.* **20**: 351–68.

Torquato, S., Hyun, S. and Donev, A. (2002). Multifunctional composites: optimizing microstructures for simultaneous transport of heat and electricity. *Phy. Rev. Lett.* **89**(26): 266601-1–4.

Xie, Y.M. and Steven, G.P. (1992). Shape and layout optimisation via an evolutionary procedure. Proceedings of International Conference on Computational Engineering Science, Hong Kong, p. 471.

Xie, Y.M. and Steven, G.P. (1997). *Evolutionary Structural Optimization*, London: Springer.

Zhou, S. and Li, Q. (2008). Design of graded two-phase microstructures for tailored elasticity gradients. *J. Mater. Sci.* **43**: 5157–67.

2

Evolutionary Structural Optimization Method

2.1 Introduction

Topology optimization may greatly enhance the performance of structures for many engineering applications. It has been exhaustively studied and various topology optimization methods have been developed over the past few decades. Among them, the evolutionary structural optimization (ESO) method is one of the most popular techniques for topology optimization (Xie and Steven 1992; Xie and Steven 1993; Xie and Steven 1997).

The ESO method was first proposed by Xie and Steven in the early 1990s (Xie and Steven 1992) and has since been continuously developed to solve a wide range of topology optimization problems (Xie and Steven 1997). ESO is based on the simple concept of gradually removing inefficient material from a structure. Through this process, the resulting structure will evolve towards its optimal shape and topology. Theoretically, one cannot guarantee that such an evolutionary procedure would always produce the best solution. However, the ESO technique provides a useful tool for engineers and architects who are interesting in exploring structurally efficient forms and shapes during the conceptual design stage of a project. Some of the original work on ESO, which was carried out by Xie and Steven (1993) and Chu *et al.* (1996, 1997), will be presented in this chapter and the deficiencies of the early versions of the ESO algorithm will be discussed.

2.2 ESO Based on Stress Level

2.2.1 Evolutionary Procedure

The stress level in any part of a structure can be determined by conducting a finite element analysis. A reliable indicator of inefficient use of material is the low values of stress (or strain) in some parts of the structure. Ideally the stress in every part of the structure should be close to the same, safe level. This concept leads to a rejection criterion based on the local stress

Evolutionary Topology Optimization of Continuum Structures: Methods and Applications   Xiaodong Huang and Mike Xie
© 2010 John Wiley & Sons, Ltd

level, where the low-stressed material is assumed to be under-utilized and is therefore removed subsequently. The removal of material can be conveniently undertaken by deleting elements from the finite element model.

The stress level of each element is determined by comparing, for example, the von Mises stress of the element σ_e^{vm} with the maximum von Mises stress of the whole structure σ_{max}^{vm}. After each finite element analysis, elements which satisfy the following condition are deleted from the model.

$$\frac{\sigma_e^{vm}}{\sigma_{max}^{vm}} < RR_i \tag{2.1}$$

where RR_i is the current rejection ratio (RR).

Such a cycle of finite element analysis and element removal is repeated using the same value of RR_i until a steady state is reached, which means that there are no more elements being deleted using the current rejection ratio. At this stage an evolutionary rate, ER, is added to the rejection ratio, i.e.

$$RR_{i+1} = RR_i + ER \tag{2.2}$$

With the increased rejection ratio the iteration takes place again until a new steady state is reached.

The evolutionary process continues until a desired optimum is obtained, for example, when there is no material in the final structure which has a stress level less than 25 % of the maximum. The evolutionary procedure can be summarized as follows:

Step 1: Discretize the structure using a fine mesh of finite elements;
Step 2: Carry out finite element analysis for the structure;
Step 3: Remove elements which satisfy the condition in (2.1);
Step 4: Increase the rejection ratio according to Equation (2.2) if a steady state is reached;
Step 5: Repeat Steps 2 to 4 until a desired optimum is obtained.

2.2.2   Example of Two-bar Frame

To find the optimal design for the loading and support conditions shown in Figure 2.1, a plane stress model with width 10 m, height 24 m and thickness 0.001 m was employed by Xie and Steven (1993). A shear stress of 1 MPa is applied on the edge of two elements at the centre of the right hand side. The whole design domain is divided into 25×60 four node elements. Young's modulus $E = 100$ GPa and Poisson's ratio $v = 0.3$ are assumed.

ESO starts from the full design using an initial rejection ratio $RR_0 = 1$ %. The evolutionary rate ER is also set to 1 %. The resulting topologies are shown in Figures 2.2(a–i) with each image showing a steady state for a given rejection ratio (Xie and Steven 1993). As the rejection ratio increases, more and more relatively inefficient material is removed from the structure. From the early stages of the evolution, it is clear that the structure is to evolve into a two-bar frame. The final two-bar system in Figure 2.2(i) gives $H = 2L$ which agrees well with the analytical solution.

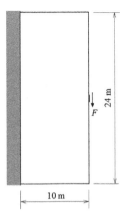

Figure 2.1 Design domain for a two-bar frame structure.

2.2.3   Examples of Michell Type Structures

The design domain for a Michell type structure with two simple supports is shown in Figure 2.3. The structure is divided into 50×25 four node plane stress elements. Young's modulus $E = 100$ GPa and Poisson's ratio $v = 0.3$ are assumed. The thickness of the plate is 0.1 m and the vertical load F is equal to 1 kN. The initial rejection ratio $RR_0 = 1\%$ and the evolutionary rate $ER = 0.5\%$ are specified.

Figures 2.4(a–e) show the ESO topologies for five steady states corresponding to rejection ratios of 5%, 10%, 15%, 20% and 25% respectively. As more and more material is removed from the structure, the stress distribution becomes more and more uniform. The ESO result at the final stage (Figure 2.4(e)) consists of an arch and four spokes between the load and the top of the arch, with the two legs of the arch being at an angle of 45° to the horizontal line. This topology bears a strong resemblance to the original Michell truss (Michell 1904).

To examine the effect of the support condition on the optimal topology, the above model is reanalysed after the simple support at the bottom right corner is replaced with a roller as shown in Figure 2.5. All other conditions and parameters remain the same as those in the example above.

Figures 2.6(a–e) show the resulting ESO designs at various stages (Xie and Steven 1993) which are clearly different from the topologies shown in Figure 2.4. There are two extra bars at the bottom to prevent the bottom right corner from moving further to the right.

2.3   ESO for Stiffness or Displacement Optimization

2.3.1   Sensitivity Number and Evolutionary Procedure

Stiffness is one of the key factors that must be taken into account in the design of such structures as buildings and bridges. Commonly the mean compliance C, the inverse measure of the overall stiffness of a structure, is considered. The mean compliance can be defined by

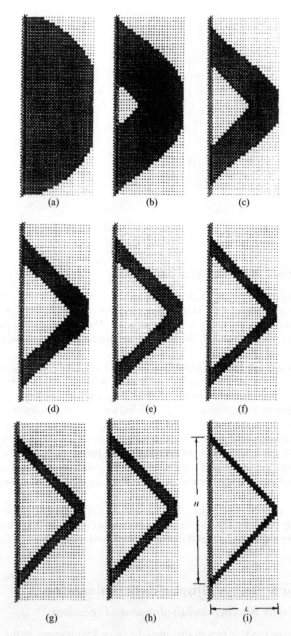

Figure 2.2 ESO topologies for a two-bar frame at different rejection ratios (Xie and Steven 1993): (a) $RR = 3\%$; (b) $RR = 6\%$; (c) $RR = 9\%$; (d) $RR = 12\%$; (e) $RR = 15\%$; (f) $RR = 18\%$; (g) $RR = 21\%$; (h) $RR = 24\%$; (i) $RR = 30\%$.

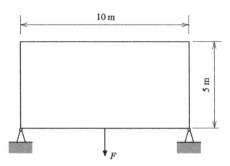

Figure 2.3   Design domain of a Michell type structure with two simple supports.

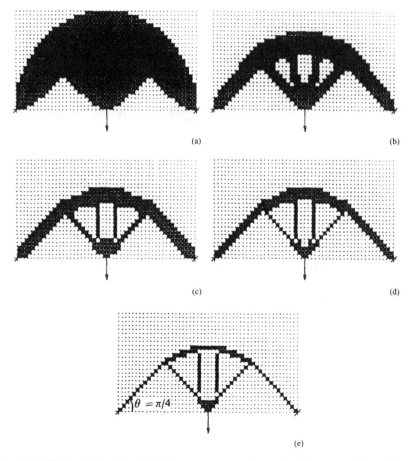

Figure 2.4   ESO topologies for a Michell type structure with two simple supports (Xie and Steven 1993): (a) $RR = 5\%$; (b) $RR = 10\%$; (c) $RR = 15\%$; (d) $RR = 20\%$; (e) $RR = 25\%$.

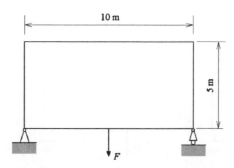

Figure 2.5 Design domain of a Michell type structure with one simple support and one roller.

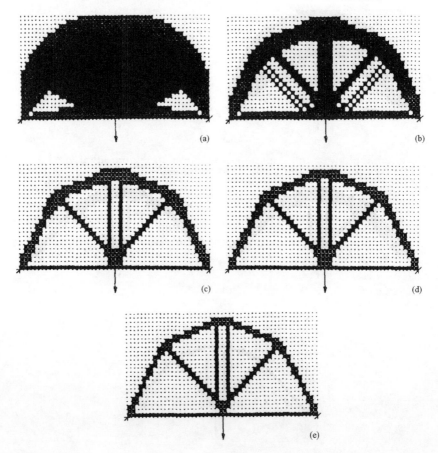

Figure 2.6 ESO topologies for a Michell type structure with one simple support and one roller (Xie and Steven 1993): (a) $RR = 5\%$; (b) $RR = 10\%$; (c) $RR = 15\%$; (d) $RR = 20\%$; (e) $RR = 25\%$.

the total strain energy of the structure or the external work done by applied loads as

$$C = \frac{1}{2}\mathbf{f}^T\mathbf{u} \qquad (2.3)$$

where $\mathbf{f}$ is the force vector and $\mathbf{u}$ is the displacement vector.

In finite element analysis, the static equilibrium equation of a structure is expressed as

$$\mathbf{Ku} = \mathbf{f} \qquad (2.4)$$

where $\mathbf{K}$ is the global stiffness matrix.

When the ith element is removed from the structure, the stiffness matrix will change by

$$\Delta\mathbf{K} = \mathbf{K}^* - \mathbf{K} = -\mathbf{K}_i \qquad (2.5)$$

where $\mathbf{K}^*$ is the stiffness matrix of the resulting structure after the element is removed and $\mathbf{K}_i$ is the stiffness matrix of the ith element. It is assumed that the removal of the element has no effect on the applied load $\mathbf{f}$. By varying both sides of Equation (2.4) and ignoring a higher order term, the change of the displacement vector is obtained as

$$\Delta\mathbf{u} = -\mathbf{K}^{-1}\Delta\mathbf{Ku} \qquad (2.6)$$

From Equations (2.3) and (2.6) we have

$$\Delta C = \frac{1}{2}\mathbf{f}^T\Delta\mathbf{u} = -\frac{1}{2}\mathbf{f}^T\mathbf{K}^{-1}\Delta\mathbf{Ku} = \frac{1}{2}\mathbf{u}_i^T\mathbf{K}_i\mathbf{u}_i \qquad (2.7)$$

where $\mathbf{u}_i$ is the displacement vector of the ith element.

Thus, the sensitivity number for the mean compliance can be defined as

$$\alpha_i^e = \frac{1}{2}\mathbf{u}_i^T\mathbf{K}_i\mathbf{u}_i \qquad (2.8)$$

The above equation indicates that the increase in the mean compliance as a result of the removal of the ith element is equal to its elemental strain energy. To minimize the mean compliance (which is equivalent to maximizing the stiffness) through the removal of elements, it is evident that the most effective way is to eliminate the elements which have the lowest values of α_i so that the increase in C will be minimal.

The number of elements to be removed is determined by the element removal ratio (ERR) which is defined as the ratio of the number of elements removed at each iteration to the total number of elements in the initial or the current FEA model (Chu et al. 1996; Chu et al. 1997).

The evolutionary procedure for stiffness optimization is given as follows:

Step 1: Discretize the structure using a fine mesh of finite elements;

Step 2: Carry out finite element analysis for the structure;

Step 3: Calculate the sensitivity number for each element using Equation (2.8);

Step 4: Remove a number of elements with the lowest sensitivity numbers according to a predefined element removal ratio ERR;

Step 5: Repeat Steps 2 to 4 until the mean compliance (or the maximum displacement, etc.) of the resulting structure reaches a prescribed limit.

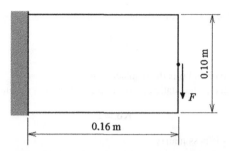

Figure 2.7   Design domain of a short cantilever.

It is worth pointing out that there is no steady state specified in the optimization for stiffness, unlike the optimization procedure based on the stress level discussed previously. If a steady state is not required, the computational efficiency can be much improved in certain cases with far less iterations. However, such a scheme may sometimes cause numerical problems in the optimization procedure, such as producing an unstable structure during an iteration.

Similar to the optimization for the overall stiffness of a structure, a sensitivity number can be derived for a displacement constraint where the maximum displacement of a structure or the displacement at a specific location has to be within a prescribed limit. Details of the sensitivity analysis and the ESO procedure for the displacement constraint have been given by Chu *et al.* (1996). A few examples are considered below.

2.3.2   Example of a Short Cantilever

The short cantilever shown in Figure 2.7 is under plane stress conditions and a vertical load of 3 kN is applied at the middle of the free end. The dimensions of the cantilever are 0.16 m (width), 0.1 m (height) and 0.001 m (thickness). Young's modulus $E = 207$ GPa and Poisson's ratio $v = 0.3$ are assumed. The optimization process will be stopped when the displacement at the load becomes greater than a prescribed value.

The design domain is divided into 32×20 four node elements. An element removal ratio of 2 % of the initial number of elements is used in each iteration. The ESO procedure creates three designs for the displacement limits of $u^* = 0.50$, 0.75 and 1.00 mm, as shown in Figures 2.8(a–c) respectively. The volumes of these designs are 56.87 %, 39.37 % and 30.00 % respectively, of the initial volume.

Next the short cantilever problem is reanalysed using different finite element meshes of 48×30 and 64×40 elements. The resulting optimal topologies are quite different, as shown in Figures 2.9(a) and (b), despite the fact that the volumes of these designs are almost the same (Chu *et al.* 1997). This clearly demonstrates that the original ESO method has failed to address the mesh-dependency problem.

2.3.3   Example of a Beam Structure

A classical topology optimization problem of a centrally loaded beam has been investigated by many researchers (Jog *et al.* 1994; Olhoff *et al.* 1991; Zhou and Rozvany 1991). The beam is

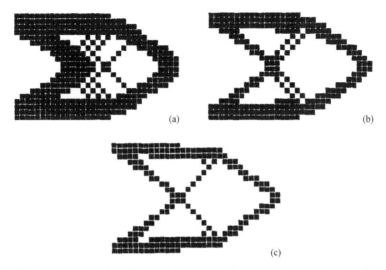

Figure 2.8 ESO topologies for stiffness optimization with various displacement limits (Chu *et al.* 1996): (a) $u^* = 0.50$ mm; (b) $u^* = 0.75$ mm; (c) $u^* = 1.00$ mm.

2400 mm long and 400 mm deep with a point load of 20 kN acting at the top middle as shown in Figure 2.10. Young's modulus $E = 200$ GPa and Poisson's ratio $v = 0.3$ are assumed. The objective is to maximize the stiffness (i.e. to minimize the compliance). The maximum deflection at the loaded point is required not to exceed 9.4 mm.

Chu *et al.* (1996; 1997) have investigated this problem using the ESO method. Because of symmetry, only half of the structure is modelled with 60×20 four node plane stress elements. The final ESO topologies using different element removal ratios are shown in Figures 2.11(a–c). The volume fractions of these optimal designs are 50.33 %, 52.42 % and 54.25 % for $ERR = 1$ %, 2 % and 4 % respectively. It is noted that a smaller element removal ratio leads to a lighter design. This is hardly surprising, but the 'better' design is obtained at a higher computational cost.

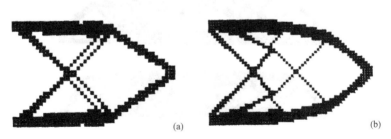

Figure 2.9 ESO topologies for stiffness optimization of a short cantilever using different mesh sizes (Chu *et al.* 1997): (a) 48×30; (b) 64×40.

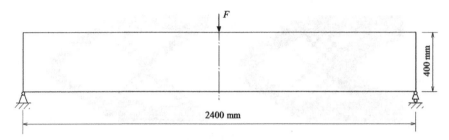

Figure 2.10   Design domain for a centrally loaded beam.

2.4   Conclusion

The original ESO method starts from the full design and removes inefficient material from a structure according to the stress or strain energy levels of the elements. The concept of ESO is very simple and can be easily understood by the user. It does not require sophisticated mathematical programming techniques. It can be readily implemented and linked to commercially available finite element analysis (FEA) software packages. No access to the source code of the FEA software is necessary. The resulting design provides a clear definition of the topology (with no "grey" area). Although the examples shown here are all plane stress problems, the ESO algorithm applies equally to general 2D and 3D problems.

The ESO method does not require regenerating new finite element meshes even when the final structure has departed substantially from the initial design. Element removal can be done by simply assigning the material property number of the rejected elements to zero and ignoring those elements when the global stiffness matrix is assembled in the subsequent finite element analysis. As more and more elements are removed, the number of equations to be solved

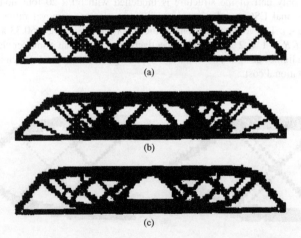

Figure 2.11   ESO topologies for stiffness optimization of a centrally loaded beam using different element removal ratio (Chu *et al.* 1997): (a) *ERR* = 1 %; (b) *ERR* = 2 %; (c) *ERR* = 4 %.

at later iterations diminishes. Therefore, a significant reduction in computation time can be achieved, especially for large 3D structures.

The ESO method has undergone a continuous development since it was proposed in 1992. Its use has been extended to topology optimization of structures with such constraints as buckling load (Manickarajah *et al.* 1998), frequency (Xie and Steven 1996), temperature (Li *et al.* 2004) or a combination of the above (Proos *et al.* 2001). The ESO technique has also been used for various engineering applications such as the underground excavation (Ren *et al.* 2005).

To minimize the material usage under a given performance constraint, the ESO method seems to follow a logical procedure to reduce the structural weight (or volume) by gradually removing material until the constraint can no longer be satisfied. However, it is possible that the material removed in an early iteration might be required later to be part of the optimal design. The ESO algorithm is unable to recover the material once it has been prematurely or wrongly deleted from the structure. Hence, while the ESO method is capable of producing a much improved solution over an initial guess design in most cases, the result may not necessarily be the absolute optimum.

The ESO algorithm is largely based on a heuristic concept and is lacking in theoretical rigour. Most of the early work on ESO neglected important numerical problems in topology optimization, such as existence of solution, checker-board, mesh-dependency and local optimum, etc. As an exception, Li *et al.* (2001) solved the checker-board problem by averaging the sensitivity number of an element with those of the neighbouring elements. To overcome the deficiencies of the ESO method, a much improved algorithm known as bi-directional evolutionary structural optimization (BESO) has been developed (Huang and Xie 2007). This will be introduced in the next chapter.

References

Chu, D.N., Xie, Y.M., Hira, A. and Steven, G.P. (1996). Evolutionary structural optimization for problems with stiffness constraints. *Finite Elements in Analysis & Design* **21**: 239–51.

Chu, D.N., Xie, Y.M., Hira, A. and Steven, G.P. (1997). On various aspects of evolutionary structrual optimization for problems with stiffness constraints. *Finite Elements in Analysis & Design* **24**: 197–212.

Huang, X. and Xie, Y.M. (2007). Convergent and mesh-independent solutions for bi-directional evolutionary structural optimization method. *Finite Elements in Analysis & Design* **43**(14): 1039–49.

Jog, C., Harber, R.B. and Bendsøe, M.P. (1994). Variable-topology shape optimization with a constraint on the perimeter. Proceedings of 20th ASME Design Automation Conf., September 11–14, 1994, Minneapolis, ASME publication DE-Vol. 69-2, 261–72.

Li, Q., Steven, G.P. and Xie, Y.M. (2001). A simple checkerboard suppression algorithm for evolutionary structural optimization. *Struct. Multidisc. Optim.* **22**: 230–9.

Li, Q., Steven, G.P., Xie, Y.M. and Querin, O.M. (2004). Evolutionary topology optimization for temperature reduction of heat conducting fields. *Inter. J. Heat & Mass Transfer* **47**: 5071–83.

Manickarajah, D., Xie, Y.M. and Steven, G.P. (1998). An evolutionary method for optimization of plate buckling resistance. *Finite Elements in Analysis and Design* **29**: 205–30.

Michell, A.G.M. (1904). The limits of economy of material in frame-structures. *Phil. Mag.* **8**: 589–97.

Olhoff, N., Bendsøe, M.P. and Rasmussen, J. (1991). On CAD-integrated structural topology and design optimization. *Comput. Meth. Appl. Mech. Engng.* **89**: 259–79.

Proos, K.A., Steven, G.P., Querin, O.M. and Xie, Y.M. (2001). Stiffness and inertia multicriteria evolutionary structural optimization. *Engng. Comput.* **18**: 1031–54.

Ren, G., Smith, J.V., Tang, J.W. and Xie, Y.M. (2005). Underground excavation shape optimization using an evolutionary procedure. *Comput & Geotech.* **32**: 122–32.

Xie, Y.M. and Steven, G.P. (1992). Shape and layout optimisation via an evolutionary procedure. *Proceedings of International Conference on Computational Engingeering Science, Hong Kong*, p. 471.

Xie, Y.M. and Steven, G.P. (1993). A simple evolutionary procedure for structural optimization. *Computers & Structures* **49**: 885–96.

Xie, Y.M. and Steven, G.P. (1996). Evolutionary structural optimization for dynamic problems. *Comput. Struct.* **58**: 1067–73.

Xie, Y.M. and Steven, G.P. (1997). *Evolutionary Structural Optimization*, London: Springer.

Zhou, M. and Rozvany, G.I.N. (1991). The COC algorithm, Part II: topological, geometrical and generalized shape optimization. *Comput. Meth. Appl. Mech. Engng.* **89**: 309–36.

3

Bi-directional Evolutionary Structural Optimization Method

3.1 Introduction

The bi-directional evolutionary structural optimization (BESO) method allows material to be removed and added simultaneously. The initial research on BESO was conducted by Yang *et al.* (1999) for stiffness optimization. In their study, the sensitivity numbers of the void elements are estimated through a linear extrapolation of the displacement field after the finite element analysis. Then, the solid elements with the lowest sensitivity numbers are removed from the structure, and the void elements with the highest sensitivity numbers are changed into solid elements. The numbers of removed and added elements in each iteration are determined by two unrelated parameters: the rejection ratio (*RR*) and the inclusion ratio (*IR*) respectively.

The BESO concept has also been applied to 'full stress design' by using the von Mises stress criterion (Querin *et al.* 2000). In their BESO algorithm, elements with the lowest von Mises stresses are removed and void elements near the highest von Mises stress regions are switched on as solid elements. Similarly, the numbers of elements to be removed and added are treated separately with a rejection ratio and an inclusion ratio respectively.

Such kind of treatment of ranking elements for removal and those for addition separately is rather cumbersome and illogical. It is noted that the user must carefully select the values of *RR* and *IR* in order to obtain a good design; otherwise, the algorithm may not produce an optimal solution (Rozvany 2009). Another problem of the early versions of BESO is that the computational efficiency is quite low because of the large number of iterations usually involved. In many cases, the final design needs to be selected from a plethora of generated topologies, and the convergence history of the objective function is often highly chaotic.

It should be pointed out that theoretically the 0/1 topology optimization problem lacks existence of solutions in its general continuum setting. The reason is that, with different mesh sizes, the introduction of more holes without changing the structural volume will generally increase the efficiency of a given design (Sigmund and Petersson 1998; Bendsøe and Sigmund 2003). This effect is regarded as a numerical instability where a larger number of holes appear when a finer finite element mesh is employed and it is termed as mesh-dependency.

Evolutionary Topology Optimization of Continuum Structures: Methods and Applications   Xiaodong Huang and Mike Xie
© 2010 John Wiley & Sons, Ltd

In this chapter, we will present a new BESO algorithm for stiffness optimization developed by authors (Huang and Xie 2007) which addresses many issues related to topology optimization of continuum structures such as a proper statement of the optimization problem, checkerboard pattern, mesh-dependency and convergence of solution.

3.2   Problem Statement and Sensitivity Number

3.2.1   Problem Statement

It is noted that the objective and constraints in the original ESO/BESO methods are very vague, especially for stiffness optimization. The original ESO appears to search for the minimum material volume subject to a given mean compliance or displacement. However, ESO may evolve into a solution much worse than the optimum if an inappropriate constraint or a large element removal ratio is applied. Tanskanen (2002) has assumed that the ESO method minimizes the compliance-volume product of a structure. However, the extended optimality of the compliance-volume product is identified to be volume fraction tending to zero or unity for most continuum structures (Rozvany *et al.* 2002). To avoid a worse design, the compliance-volume product is used as a performance index (Liang *et al.* 2000) and the evolutionary procedure is stopped once the performance index drops dramatically. However, the problem may still arise – for example, the whole evolutionary procedure may have been stopped *before* a prescribed displacement constraint is satisfied.

Topology optimization is often aimed at searching for the stiffest structure with a given volume of material. In ESO/BESO methods, a structure is optimized by removing and adding elements. That is to say that, the element itself, rather than its associated physical or material parameters, is treated as the design variable. Thus, the optimization problem with the volume constraint is stated as

$$\text{Minimize} \quad C = \frac{1}{2}\mathbf{f}^{\mathrm{T}}\mathbf{u} \tag{3.1a}$$

$$\text{Subject to}: \quad V^* - \sum_{i=1}^{N} V_i x_i = 0 \tag{3.1b}$$

$$x_i = 0 \quad \text{or} \quad 1 \tag{3.1c}$$

where $\mathbf{f}$ and $\mathbf{u}$ are the applied load and displacement vectors and C is known as the mean compliance. V_i is the volume of an individual element and V^* the prescribed total structural volume. N is the total number of elements in the system. The binary design variable x_i declares the absence (0) or presence (1) of an element.

The above problem statement has been widely used for the topology optimization of continuum structure (Bendsøe and Sigmund 2003) but differs from the one used in the original ESO/BESO methods. In fact, the original ESO/BESO methods have difficulties in dealing with the problem stated in Equations (3.1a–c) – for example, the objective function may not converge if the volume is kept constant to satisfy the volume constraint in Equation (3.1b). One of the aims of the new BESO method is to make the algorithm to be stably convergent towards a solution which exactly addresses the above optimization problem statement.

3.2.2   Sensitivity Number

When a solid element is removed from a structure, the change of the mean compliance or total strain energy is equal to the elemental strain energy (Chu *et al.* 1996). This change is defined as the elemental sensitivity number:

$$\alpha_i^e = \Delta C_i = \frac{1}{2}\mathbf{u}_i^T\mathbf{K}_i\mathbf{u}_i \qquad (3.2)$$

where $\mathbf{u}_i$ is the nodal displacement vector of the ith element, $\mathbf{K}_i$ is the elemental stiffness matrix. When a nonuniform mesh is assigned, the sensitivity number should consider the effect of the volume of the element. In such a case, the sensitivity number can be replaced with the elemental strain energy density as

$$\alpha_i^e = e_i = \left(\frac{1}{2}\mathbf{u}_i^T\mathbf{K}_i\mathbf{u}_i\right)\Big/\mathrm{V}_i \qquad (3.3)$$

The original ESO procedure for stiffness optimization is directly driven by gradually removing elements with the lowest sensitivity numbers defined in Equation (3.2) or (3.3).

The sensitivity numbers for void elements are assumed to be zero initially. The reason for this setting will be given in the next chapter. To add material into the design domain, a filter scheme will be used to obtain the sensitivity number for the void elements and to smooth the sensitivity number in the whole design domain. More importantly, by using the filter scheme the problems of checkerboard pattern and mesh-dependency will be resolved at once.

3.3   Filter Scheme and Improved Sensitivity Number

3.3.1   Checkerboard and Mesh-dependency Problems

When a continuum structure is discretized using low order bilinear (2D) or trilinear (3D) finite elements, the sensitivity numbers could become C^0 discontinuous across element boundaries. This leads to checkerboard patterns in the resulting topologies (Jog and Harber 1996). Figure 3.1 shows a typical checkerboard pattern of a continuum structure from the original ESO method. The presence of checkerboard pattern causes difficulty in interpreting and manufacturing the 'optimal' structure. To suppress the formation of checkerboard patterns in the ESO method, a simple smoothing scheme of averaging the sensitivity numbers of neighbouring elements has been presented by Li *et al.* (2001). However, this smoothing algorithm cannot overcome the mesh-dependency problem.

The so-called mesh-dependency refers to the problem of obtaining different topologies from using different finite element meshes. When a finer mesh is used, the numerical process of structural optimization will produce a topology with more members of smaller sizes in the final design. Ideally, mesh-refinement should result in a better finite element modelling of the same optimal structure and a better description of boundaries – not in a more detailed or qualitatively different structure (Bendsøe and Sigmund 2003).

Various techniques have been suggested to overcome the mesh-dependency problem such as the perimeter control method (Harber *et al.* 1996; Jog 2002) and the sensitivity filter scheme (Sigmund 1997; Sigmund and Petersson 1998). The BESO method with perimeter control (Yang *et al.* 2003) is demonstrated to be capable of obtaining mesh-independent solutions

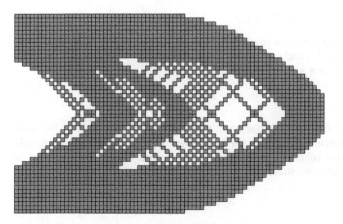

Figure 3.1   A typical checkerboard pattern in the ESO method.

because of one extra constraint (the perimeter length) on the topology optimization problem. However, predicting or selecting an appropriate value of the perimeter length for a new design problem can be a difficult task. Thus, the sensitivity filter scheme will be introduced into the new BESO method.

3.3.2   Filter Scheme for BESO Method

Before applying the filter scheme, nodal sensitivity numbers which do not carry any physical meaning on their own are defined by averaging the elemental sensitivity numbers as follows:

$$\alpha_j^n = \sum_{i=1}^{M} w_i \alpha_i^e \tag{3.4}$$

where M denotes the total number of elements connected to the jth node. w_i is the weight factor of the ith element and $\sum_{i=1}^{M} w_i = 1$. w_i can be defined by

$$w_i = \frac{1}{M-1} \left(1 - \frac{r_{ij}}{\sum_{i=1}^{M} r_{ij}} \right) \tag{3.5}$$

where r_{ij} is the distance between the centre of the ith element and the jth node. The above weight factor indicates that the elemental sensitivity number has larger effect on the nodal sensitivity number when it is closer to the node.

The above nodal sensitivity numbers will then be converted into smoothed elemental sensitivity numbers. This conversion takes place through projecting nodal sensitivity numbers to the design domain. Here, a filter scheme is used to carry out this process. The filter has a length scale r_{min} that does not change with mesh refinement. The primary role of the scale parameter

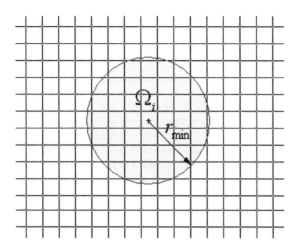

Figure 3.2 Nodes located inside the circular sub-domain Ω_i are used in the filter scheme for the ith element.

$r_{\min}$ in the filter scheme is to identify the nodes that will influence the sensitivity of the ith element. This can be visualized by drawing a circle of radius $r_{\min}$ centred at the centroid of ith element, thus generating the circular sub-domain Ω_i as shown in Figure 3.2. Usually the value of $r_{\min}$ should be big enough so that Ω_i covers more than one element. The size of the sub-domain Ω_i does not change with mesh size. Nodes located inside Ω_i contribute to the computation of the improved sensitivity number of the ith element as

$$\alpha_i = \frac{\sum_{j=1}^{K} w(r_{ij}) \alpha_j^n}{\sum_{j=1}^{K} w(r_{ij})} \tag{3.6}$$

where K is the total number of nodes in the sub-domain Ω_i, $w(r_{ij})$ is the linear weight factor defined as

$$w(r_{ij}) = r_{\min} - r_{ij} \quad (j = 1, 2, \ldots, K) \tag{3.7}$$

It can be seen that the filter scheme smoothes the sensitivity numbers in the whole design domain. Thus, the sensitivity numbers for void elements are automatically obtained. They may have high values due to high sensitivity numbers of solid elements within the sub-domain Ω_i. Therefore, some of the void elements may be changed to solid elements in the next iteration.

The above filter scheme is similar to the mesh-independency filter used by Sigmund and Petersson (1998) except that Equation (3.6) uses the nodal sensitivity numbers rather than elemental sensitivities. It is noted that Sigmund and Petersson (1998) include the density of the element in the filter, thus the sensitivity number for void elements will be infinite.

However, the present filter given in Equation (3.6) does not consider the element status (void or solid), and the initial sensitivity number for void elements is set to zero. Using the above filter technique, nonzero sensitivity numbers for void elements are obtained as a result of filtering the sensitivity numbers of neighbouring solid elements. Using the obtained sensitivity numbers, the void elements can be ranked alongside the solid elements in terms of their structural importance.

The filter scheme is purely heuristic. However, by adopting this simple technique, many numerical problems in topology optimization, such as checkerboard and mesh-dependency, can be effectively overcome. It produces results very similar to those obtained by applying a local gradient constraint (Bendsøe and Sigmund 2003). The filter scheme requires little extra computational time and is very easy to implement in the optimization algorithm.

3.3.3   Stabilizing the Evolutionary Process

As will be demonstrated later in this chapter, the adoption of the above filter scheme can effectively address the mesh-dependency problem. However, the objective function and the corresponding topology may not be convergent. Let us consider a short cantilever example similar to the one shown in Figure 2.7 using a coarse mesh of 32×20 elements. With ESO/BESO methods, large oscillations are often observed in the evolution history of the objective function, as illustrated in the Figure 3.3(a). The reason for such chaotic behaviour is that the sensitivity numbers of the solid (1) and void (0) elements are based on discrete design variables of element presence (1) and absence (0). This makes the objective function and the topology difficult to converge. Huang and Xie (2007) has found that averaging the sensitivity number with its historical information is an effective way to solve this problem. The simple averaging scheme is given as

$$\alpha_i = \frac{\alpha_i^k + \alpha_i^{k-1}}{2} \tag{3.8}$$

where k is the current iteration number. Then let $\alpha_i^k = \alpha_i$ which will be used for the next iteration. Thus, the updated sensitivity number includes the whole history of the sensitivity information in the previous iterations. Figure 3.3(b) shows the evolution history obtained by adopting the stabilization scheme defined in equation (3.8). Compared to the result in Figure 3.3(a), the new solution is highly stable in both the topology and the objective function (the mean compliance) after the constraint volume fraction (50 %) is achieved. It is worth pointing out that whilst Equation (3.8) affects the search path of the BESO algorithm it has very little effect on the final solution when it becomes convergent. Details of the parameters used in this example can be found in Huang and Xie (2007).

3.4   Element Removal/Addition and Convergence Criterion

Before elements are removed from or added to the current design, the target volume for the next iteration (V_{k+1}) needs to be given first. Since the volume constraint (V^*) can be greater or smaller than the volume of the initial guess design, the target volume in each iteration may decrease or increase step by step until the constraint volume is achieved. The evolution of the

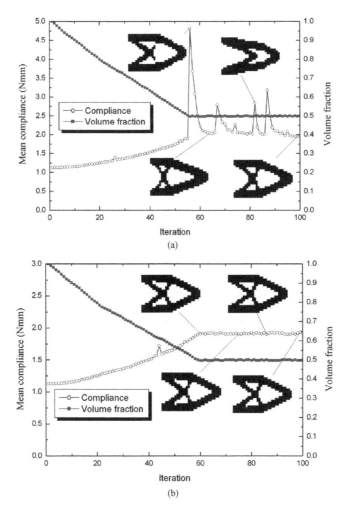

Figure 3.3   Comparison of evolution histories: (a) without the stabilization scheme; (b) with the stabilization scheme.

volume can be expressed by

$$V_{k+1} = V_k(1 \pm ER) \quad (k = 1, 2, 3 \cdots) \tag{3.9}$$

where ER is the evolutionary volume ratio. Once the volume constraint is satisfied, the volume of the structure will be kept constant for the remaining iterations as

$$V_{k+1} = V^* \tag{3.10}$$

Then the sensitivity numbers of all elements, both solid and void, are calculated as described in the previous sections. The elements are sorted according to the values of their sensitivity numbers (from the highest to the lowest). For solid element (1), it will be removed (switched to 0) if

$$\alpha_i \leq \alpha_{del}^{th} \tag{3.11}$$

For void elements (0), it will be added (switched to 1) if

$$\alpha_i > \alpha_{add}^{th} \tag{3.12}$$

where α_{del}^{th} and α_{add}^{th} are the threshold sensitivity numbers for removing and adding elements, and α_{del}^{th} is always less than or equal to α_{add}^{th}. α_{del}^{th} and α_{add}^{th} are determined by the following three simple steps:

1. Let $\alpha_{add}^{th} = \alpha_{del}^{th} = \alpha_{th}$, thus α_{th} can be easily determined by V_{k+1}. For example, there are 1000 elements in the design domain and $\alpha_1 > \alpha_2 \cdots > \alpha_{1000}$ and if V_{k+1} corresponds to a design with 725 solid elements then $\alpha_{th} = \alpha_{725}$.
2. Calculate the volume addition ratio (AR), which is defined as the number of added elements divided by the total number of elements in the design domain. If $AR \leq AR_{max}$ where AR_{max} is a prescribed maximum volume addition ratio, skip step 3. Otherwise recalculate α_{del}^{th} and α_{add}^{th} as in step 3.
3. Calculate α_{add}^{th} by first sorting the sensitivity number of void elements (0). The number of elements to be switched from 0 to 1 will be equal to AR_{max} multiplied by the total number of elements in the design domain. α_{add}^{th} is the sensitivity number of the element ranked just below the last added element. α_{del}^{th} is then determined so that the removed volume is equal to ($V_k - V_{k+1}$ + the volume of the added elements).

It is seen that AR_{max} is introduced to ensure that not too many elements are added in a single iteration. Otherwise, the structure may lose its integrity when the BESO method starts from an initial guess design. Normally AR_{max} is greater than 1 % so that it does not suppress the capability or advantages of adding elements.

It is noted that the new element removal and addition scheme ranks all elements (void and solid) together, while in the previous BESO method (Yang *et al.* 1999, Querin *et al.* 2000) elements for removal and those for addition are treated differently and ranked separated, which is cumbersome and illogical.

The cycle of finite element analysis and element removal/addition continues until the objective volume (V^*) is reached *and* the following convergence criterion (defined in terms of the change in the objective function) is satisfied.

$$error = \frac{\left| \sum_{i=1}^{N} C_{k-i+1} - \sum_{i=1}^{N} C_{k-N-i+1} \right|}{\sum_{i=1}^{N} C_{k-i+1}} \leq \tau \tag{3.13}$$

where k is the current iteration number, τ is a allowable convergence tolerance and N is an integer number. Normally, N is selected to be 5 which implies that the change in the mean compliance over the last 10 iterations is acceptably small.

3.5   Basic BESO Procedure

The evolutionary iteration procedure of the present BESO method is given as follows:

1. Discretize the design domain using a finite element mesh and assign initial property values (0 or 1) for the elements to construct an initial design.
2. Perform finite element analysis and then calculate the elemental sensitivity number according to Equation (3.6).
3. Average the sensitivity number with its history information using Equation (3.8) and then save the resulted sensitivity number for next iteration.
4. Determine the target volume for the next iteration using Equation (3.9).
5. Add and delete elements according to the procedure described in Section 3.4.
6. Repeat steps 2–5 until the constraint volume (V^*) is achieved and the convergence criterion (3.13) is satisfied.

A flowchart of the BESO method is given in Figure 3.4. The BESO algorithm is programmed in Visual Fortran and integrated with ABAQUS software that is used as the FEA solver.

3.6   Examples of BESO Starting from Initial Full Design

3.6.1   Topology Optimization of a Short Cantilever

This example considers the stiffness optimization of a short cantilever shown in Figure 3.5. The design domain is 80 mm in length, 55 mm in height and 1 mm in thickness. A 100 N downward force is applied at the centre of the free end. Young's modulus of 100 GPa and Poisson's ratio of 0.3 are assumed. The available material will cover 50 % of the design domain. In other words, the volume fraction of the final design will be 50 %. BESO starts from the full design which is subdivided using a mesh of 160×100 four node plane stress elements. The BESO parameters are: $ER = 1\,\%$, $AR_{max} = 5\,\%$, $r_{min} = 3$ mm and $\tau = 0.01\,\%$.

Figure 3.6 shows the evolution histories of the mean compliance and the volume fraction. The mean compliance increases and the topology develops gradually as material is removed from the design domain. It is worth pointing out that the occasional jumps in the mean compliance are caused by a significant change of topology resulting from the elimination of one or more bars in one iteration, as can be observed from Figure 3.7. Following a jump, the mean compliance quickly recovers and resumes a smooth ascent. Note that in order to find the stiffest structure for a given volume, the increase in the mean compliance should be kept as small as possible. It is seen from Figure 3.6 that after the volume reaches the objective volume (50 % of the full design), the mean compliance converges to an almost constant value, 1.87 Nmm. Figure 3.7 shows the evolution history of topologies at various iterations. The final solution is given in Figure 3.7(f) and the total iteration number is 79.

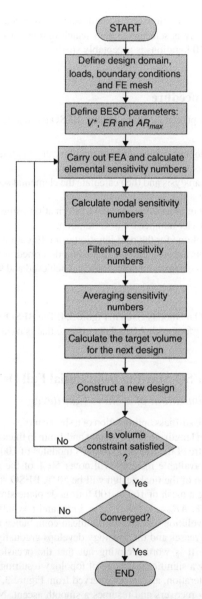

Figure 3.4   Flowchart of the BESO method.

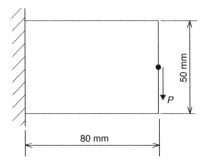

Figure 3.5 Dimensions of the design domain, and boundary and loading conditions for a short cantilever.

To verify the developed BESO method, the above problem using the same mesh is solved using the SIMP method (Sigmund 2001) with penalty factor $p = 3$ and filter radius $r_{min} = 3$ mm. The final topology is shown in Figure 3.8, which is very similar to the BESO solution in Figure 3.7(f). It is noted that the SIMP solution has 'grey' elements of intermediate material densities. Its mean compliance is 2.07 Nmm which is higher than that of the BESO topology (1.87 Nmm). The difference may be attributed to the over-estimated strain energy of elements with intermediate densities in the SIMP topology.

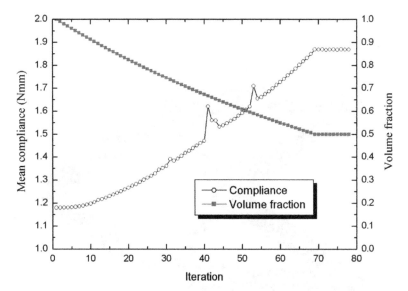

Figure 3.6 Evolution histories of mean compliance and volume fraction when BESO starts from the full design.

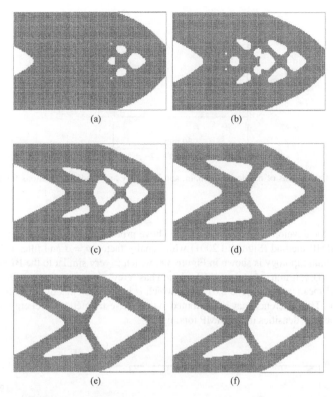

(a)   (b)

(c)   (d)

(e)   (f)

Figure 3.7   Evolution of topology: (a) iteration 15; (b) iteration 30; (c) iteration 45; (d) iteration 60; (e) iteration 69; (f) final solution (iteration 79).

Figure 3.8   Optimal topology for the cantilever using the SIMP method.

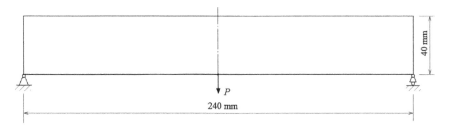

Figure 3.9 Dimensions of the design domain, and boundary and loading conditions for a beam.

3.6.2 Topology Optimization of a Beam

The beam shown in Figure 3.9 is loaded at its bottom centre by $P = 100$ N. The symmetric right half of the 120 mm × 40 mm domain (with 1 mm thickness) is discretized using 120 × 40 four node plane stress elements. Suppose only 50 % of the design domain volume is available for constructing the final structure and the material has Young's modulus $E = 200$ GPa and Poisson's ratio $v = 0.3$. Initially, the material occupies the full design domain. The BESO parameters used in this example are: $ER = 5\%$, $AR_{max} = 5\%$, $r_{min} = 6$ mm and $\tau = 0.01\%$.

Figure 3.10 shows the evolution histories of the mean compliance and the volume fraction. The mean compliance converges to a stable value at the final stage. Figure 3.11 shows the

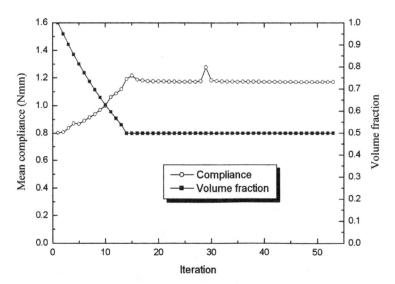

Figure 3.10 Evolution histories of mean compliance and volume fraction when BESO starts from the full design.

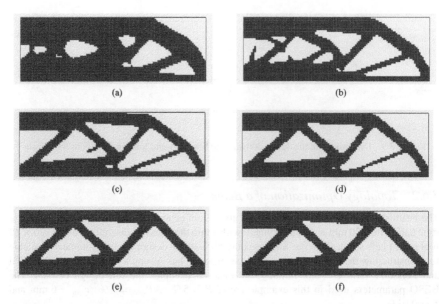

(a)   (b)

(c)   (d)

(e)   (f)

Figure 3.11   Evolution of topology: (a) iteration 5; (b) iteration 10; (c) iteration 15; (d) iteration 25; (e) iteration 40; (f) final solution (iteration 53).

evolution history of the topology. It can be seen that the topology has very little change after the 40th iteration. The final optimal design is shown in Figure 3.11(f) and its mean compliance is 1.17 Nmm. The total number of iterations is 53.

It is noted that if a more relaxed convergence criterion were used (i.e. a larger tolerance τ is employed), the solution would have been considered convergent before a jump occurred at iteration 29. A slightly bigger tolerance τ would have had no significant effect on the value of the final mean compliance, but could have resulted in quite different a topology. This is evident from comparing the result at iteration 25 (Figures 3.11(d)) with that at iteration 53 (Figure 3.11(f)).

When the above problem is solved by the SIMP method with penalty factor $p = 3$ and $r_{min} = 6$ mm, the optimal topology is shown in Figure 3.12. As expected, this topology is also

Figure 3.12   Optimal topology of the beam using the SIMP method.

similar to the BESO solution shown in Figure 3.11(f). The predicted mean compliance of the SIMP topology is 1.26 Nmm which is also higher than that of the BESO topology.

3.7   Examples of BESO Starting from Initial Guess Design

The above examples show that the BESO method can find the optimal topology by starting from the full design. However, before the constraint volume is reached the BESO procedure is in effect searching for an appropriate guess design for later iterations. In this section, we perform BESO differently by starting from an initial guess design which has a volume equal or close to the objective volume. One advantage of this approach is that only a portion of elements in the design domain is involved in the analysis and therefore the computational time could be markedly decreased, especially for a large 3D model. Compared to BESO starting from a minimum structure with very few elements, BESO starting from a guess design with its volume close to the target reduces the number of iterations for adding material to the structure.

Let us reconsider the short cantilever problem discussed in section 3.6.1 by starting BESO from the initial guess design in Figure 3.13(a). The final topology is given in Figure 3.13(b). Figure 3.14 shows the evolution histories of the mean compliance and the volume fraction. It can be seen that the mean compliance is convergent. The final topology has a mean compliance of 1.88 Nmm. Compared to the optimal design in Figure 3.7(f), which has a mean compliance of 1.87 Nmm, the topologies and the mean compliances from the two BESO approaches are very close.

For topology optimization of the beam given in Figure 3.9, the initial guess design and the final topology are shown in Figure 3.15. The evolution histories of the mean compliance and the volume fraction are shown in Figure 3.16. Again the mean compliance is convergent. The final topology has a mean compliance of 1.18 Nmm which is very close to that of the topology shown in Figure 3.11(f).

These calculations show that the present BESO method always leads to a convergent solution even if the guess design is totally different from the final topology. The examples also demonstrated that the BESO starting from the full design can find the optimum even when the search direction occasionally deviates from the correct path. However, BESO starting from an initial guess design may sometimes converge to a local optimum because some void elements in the initial guess design may never be included in the finite element analysis during the whole optimization process. To eliminate or reduce the likelihood of a local optimum, it might

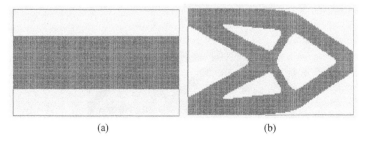

(a)   (b)

Figure 3.13   BESO starts from an initial guess design: (a) initial guess design; (b) final topology.

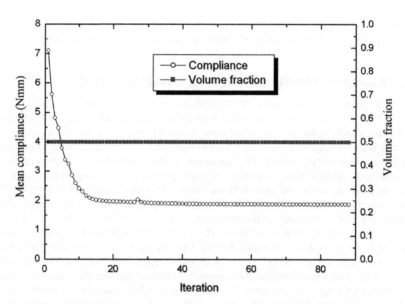

Figure 3.14   Evolution histories of mean compliance and volume fraction when BESO starts from an initial guess design.

be necessary that BESO should start from the full design so that all elements are involved in the finite element analysis at least once.

In terms of the computational time, BESO starting from an initial guess design has not led to significant savings for above two examples because a large number of iterations is required to evolve a guess design to the optimum especially when the guess design is far different from the optimal topology. However, as is evident from Figures 3.14 and 3.16, it is possible to get substantial savings in computational time if a larger convergence tolerance τ is used so that the

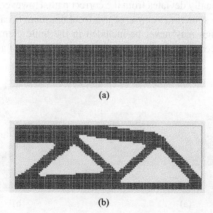

Figure 3.15   BESO starts from an initial guess design: (a) initial guess design; (b) final topology.

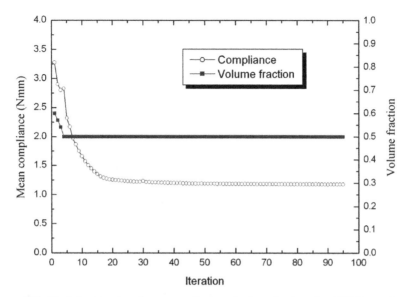

Figure 3.16 Evolution histories of mean compliance and volume fraction when BESO starts from an initial guess design.

solution would be considered convergent in less than 30 iterations. In doing so, a design with its objective function slightly higher (worse) than that of the 'optimum' would be obtained. As has been mentioned before, sometimes even if the difference in the values of the objective function is small, the corresponding topologies could be quite different.

3.8   Example of a 3D Structure

A 3D example is given here to demonstrate the computational efficiency of the present BESO method. Figure 3.17 shows the design domain of a 3D cantilever with a concentrated load $F = 1$ kN acting at the centre of the free end. Because of symmetry, only half of the structure is modelled using total 40 000 eight node brick elements. Young's modulus $E = 10$ GPa and Poisson's ratio $v = 0.3$ are assumed. The objective volume is 10 % of the total volume of the design domain. BESO starts from the full design. Other BESO parameters are: $ER = 3$ %, $AR_{max} = 50$ %, $r_{min} = 3$ mm and $\tau = 0.1$ %. It is noted that a very large AR_{max} is adopted which in effect removes the limit on the number of elements allowed to be added in each iteration.

Figure 3.18 shows the evolution histories of the mean compliance and the volume fraction and Figure 3.19 gives the resulting topologies at various iterations. Generally the mean compliance increases as the volume of the structure gradually decreases. A big jump in the mean compliance has occurred at iteration 75 due to the breakage of some bars. Afterwards, the mean compliance gradually decreases while the volume fraction is kept constant at 10 %. At the final stage, both the mean compliance and the topology (see Figures 3.19(e) and (f)) become convergent. The mean compliance of the final optimal design (Figure 3.19(f)) is 1863 Nmm.

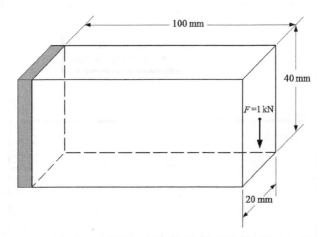

Figure 3.17   Design domain for a 3D cantilever.

The whole optimization process takes 2 hours and 40 minutes on a Personal Computer with Pentium 4 3.0 GHz processor and 512 MB RAM. The detailed computational time for each iteration is shown in Figure 3.20. It is seen that the computational time for each iteration becomes shorter and shorter (from 319 seconds for the first iteration to 57 seconds for the last iteration) as a result of less and less elements in the finite element model.

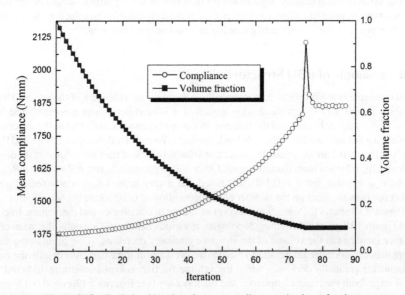

Figure 3.18   Evolution histories of mean compliance and volume fraction.

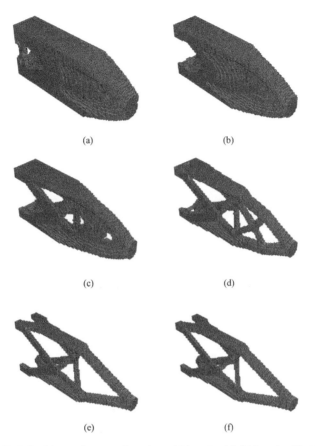

(a)   (b)

(c)   (d)

(e)   (f)

Figure 3.19 Evolution history of structural topology: (a) iteration 15; (b) iteration 30; (c) iteration 45; (d) iteration 60; (e) iteration 80; (f) iteration 87.

3.9 Mesh-independence Studies

The filter scheme presented in Section 3.3.2 is based on an image-processing technique and works as a low-pass filter that eliminates structural components below a certain length-scale in the optimal design. One of the benefits of adopting such a filter scheme is that the optimal topology will no longer be dependent on the mesh sizes. This will be demonstrated below using two examples.

First the short cantilever example shown in Figure 3.5 is optimized using four different mesh sizes: 32×20, 80×50, 160×100, 240×150, with the same filter radius $r_{min} = 3$ mm. Figure 3.21 gives the optimal topologies obtained from the different mesh sizes. It clearly shows that despite the significant differences in the mesh sizes the optimal topologies remain largely the same. The only difference in the topologies is that the boundary of the structure becomes smoother with mesh refinement.

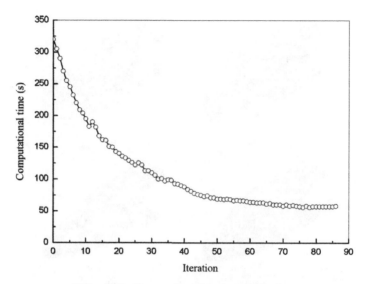

Figure 3.20   Computational time for each iteration.

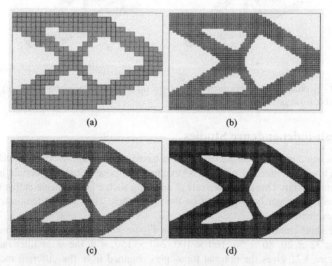

Figure 3.21   Mesh-independent solutions for the cantilever from different mesh sizes: (a) 32 × 20; (b) 80 × 50; (c) 160 × 100; (d) 240 × 150.

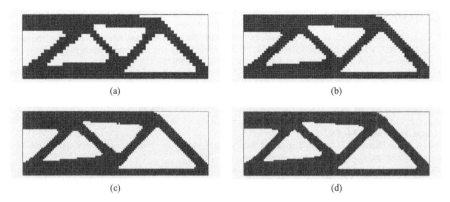

Figure 3.22 Mesh-independent solutions for the beam from different mesh sizes: (a) 60×20; (b) 90×30; (c) 120×40; (d) 150×50.

Next, we solve the beam optimization problem with four different mesh sizes: 60×20, 90×30, 120×40, 150×50. The optimal topologies are shown in Figure 3.22. Again, mesh-independent designs are obtained by adopting the proposed filter scheme for BESO.

3.10   Conclusion

We have presented in this chapter a new BESO method for stiffness optimization. The sensitivity numbers of elements are based on the elemental strain energy density. The sensitivity numbers used for material removal and addition are modified by introducing a mesh-independency filter which smoothes the sensitivity numbers throughout the design domain. Also, the convergence histories of the mean compliance and the structural topology are greatly improved by averaging the sensitivity numbers with their historical information.

Examples have shown the capability of the present BESO method to obtain convergent, checkerboard-free and mesh-independent solutions. The resulted topologies compare very well with those of the SIMP method. Furthermore, the BESO method may start from guess designs that are much smaller than the full design domain and may therefore save computational time for the finite element analysis.

The current BESO method is often called 'hard-kill' BESO method due to the complete removal of an element (as opposed to changing it into a very soft material). The main advantage of the hard-kill approach is that the computational time can be significantly reduced, especially for large 3D structures, because the eliminated elements are not involved in the finite element analysis. However, there have been some doubts among researchers about the theoretical correctness of the hard-kill ESO/BESO methods, especially after Zhou and Rozvany (2001) has showed that ESO fails on a certain problem. To gain a deeper understanding of the BESO method, we shall discuss an artificial material interpolation scheme for BESO in the next chapter and then present a comparison between BESO and various other topology optimization methods in Chapter 5.

References

Bendsøe, M.P. and Sigmund, O. (2003). *Topology Optimization: Theory, Method and Application*. Berlin: Springer.

Chu, D.N., Xie, Y.M., Hira, A. and Steven, G.P. (1996). Evolutionary structural optimization for problems with stiffness constraints. *Finite Elements in Analysis and Design* **21**: 239–51.

Harber, R.B., Jog, C.S. and Bendsøe, M.P. (1996). A new approach to variable-topology shape design using a constraint on the perimeter. *Struct. Optim.* **11**: 1–11.

Huang, X. and Xie, Y.M. (2007). Convergent and mesh-independent solutions for bi-directional evolutionary structural optimization method. *Finite Elements in Analysis and Design* **43**(14): 1039–49.

Jog, C.S. (2002). Topology design of structures using a dual algorithm and a constraint on the perimeter. *Inter. J. Num. Meth .Engng.* **54**: 1007–19.

Jog, C.S. and Harber, R.B. (1996). Stability of finite element models for distributed-parameter optimization and topology design. *Comput. Meth. Appl. Mech. Engng.* **130**: 1951–65.

Li, Q., Steven, G.P. and Xie, Y.M. (2001). A simple checkerboard suppression algorithm for evolutionary structural optimization. *Struct. Multidisc. Optim.* **22**: 230–9.

Liang, Q.Q., Xie, Y.M. and Steven, G.P. (2000). Optimal topology selection of continuum structures with displacement constraints. *Comput. & Struct.* **77**: 635–44.

Querin, O.M., Young, V., Steven, G.P. and Xie, Y.M. (2000). Computational efficiency and validation of bi-directional evolutionary structural optimization. *Comput. Meth. Appl. Mech. Engng.* **189**: 559–73.

Rozvany, G.I.N. (2009). A critical review of established methods of structrual topology optimization. *Struct. Multidisc. Optim.* **37**(3): 217–37.

Rozvany, G.I.N., Querin, O.M., Gaspar, Z. and Pomezanski, V. (2002). Extended optimality in topology design. *Struct. Multidisc. Optim.* **24**: 257–61.

Sigmund, O. (1997). On the design of compliant mechanisms using topology optimization. *Mech. Struct. Mach.* **25**: 495–526.

Sigmund, O. (2001). A 99 line topology optimization code written in MATLAB®. *Struct. Multidisc. Optim.* **21**: 120–7.

Sigmund, O. and Petersson, J. (1998). Numerical instabilities in topology optimization: A survey on procedures dealing with checkerboards, mesh-dependencies and local minima. *Struct. Optim.* **16**: 68–75.

Tanskanen, P. (2002). The evolutionary structural optimization method: theoretical aspects. *Comput. Meth. Appl. Mech. Engng.* **191**: 5485–98.

Yang, X.Y., Xie, Y.M., Liu, G.T. and Clarkson, P.J. (2003). Perimeter control of the bidirectional evolutionary optimization method. *Struct. Multidisc. Optim.* **24**: 430–40.

Yang, X.Y., Xie, Y.M., Steven, G.P. and Querin, O.M. (1999). Bidirectional evolutionary method for stiffness optimization. *AIAA Journal* **37**: 1483–8.

Zhou, M. and Rozvany, G.I.N. (2001). On the validity of ESO type methods in topology optimization. *Struct. Multidisc. Optim.* **21**: 80–3.

4

BESO Utilizing Material Interpolation Scheme with Penalization

4.1 Introduction

In the BESO method, the optimal topology is determined according to the relative ranking of sensitivity numbers. The sensitivity numbers of the solid elements can be easily estimated by the approximate variation of the objective function due to the removal of individual elements. However, it is difficult to estimate the sensitivity numbers of the void elements because there is hardly any information available for the void elements which are not included in the finite element analysis. In the previous chapter, sensitivity numbers of void elements were set to be zero initially and then modified through the filter scheme. It is shown that by applying the filter scheme and incorporating the historical information in the sensitivity numbers, the hard-kill BESO algorithm leads to a convergent and mesh-independent solution, and the resulting topology is very similar to the optimal topology of the SIMP method.

However, the complete removal of a solid element from the design domain could result in theoretical difficulties in topology optimization. It appears to be rather irrational when the design variable (an element) is directly eliminated from the topology optimization problem. An alterative way of effectively 'removing' an element is to reduce the elastic modulus of the element or one of the characteristic dimensions (such as the thickness) of the element to a very small value. For example, Hinton and Sienz (1995) reduce the elastic modulus of elements which are to be 'removed' by dividing a factor 10^5. Rozvany and Querin (2002) suggest a sequential element rejection and admission (SERA) method in which the void element is replaced by a soft element with a very low density. A similar approach for BESO has been proposed by Zhu et al. (2007) where an orthotropic cellular microstructure is introduced to replace the void element. However, none of these methods produces significantly different a topology from that of the original ESO method.

The sensitivity number in the original ESO/BESO methods for stiffness optimization is defined by the approximate variation of the objective function due to removing an element.

Evolutionary Topology Optimization of Continuum Structures: Methods and Applications   Xiaodong Huang and Mike Xie
© 2010 John Wiley & Sons, Ltd

As shown in Chapter 2, the derivation of the sensitivity number differs from the conventional sensitivity analysis in structural optimization. A thorough examination of the relationship between the sensitivity number and the conventional sensitivity analysis is warranted because of its importance for further extending the BESO method to other topology optimization problems.

Similarities in the resulting topologies from BESO and SIMP may well indicate some intrinsic relationships between the sensitivity number of BESO and the sensitivity analysis in the SIMP material model. In this chapter we introduce a soft-kill BESO method utilizing the material interpolation scheme with penalization (Huang and Xie 2009).

4.2   Problem Statement and Material Interpolation Scheme

4.2.1   Problem Statement

To obtain a solid-void design with maximum stiffness, the mean compliance is minimized for a given volume of material. The topology optimization problem can be stated as

$$\text{Minimize } C = \frac{1}{2}\mathbf{f}^\mathsf{T}\mathbf{u} \tag{4.1a}$$

$$\text{Subject to}: V^* - \sum_{i=1}^{N} V_i x_i = 0 \tag{4.1b}$$

$$x_i = x_{\min} \text{ or } 1 \tag{4.1c}$$

where V_i is the volume of an individual element and V^* the prescribed volume of the final structure. The binary design variable x_i denotes the relative density of the ith element. It can be seen that the only difference between Equation (4.1) and Equation (3.1) is that a small value of $x_{\min}$ (e.g. 0.001) is used in the above problem statement for the void element. This indicates that no element is allowed to be completely removed from the design domain. Such an approach is classified as a soft-kill method.

4.2.2   Material Interpolation Scheme

Material interpolation schemes with penalization have been widely used in the SIMP method to steer the solution to nearly solid-void designs (Bendsøe 1989; Bendsøe and Sigmund 2003; Rietz 2001; Zhou and Rozvany 1991). Bendsøe and Sigmund (1999) have compared the material interpolation schemes to various bounds for effective material properties in composite (e.g. the Hashin-Shtrikman bounds) and showed that composite materials from intermediate densities are physically realizable. To achieve a nearly solid-void design, Young's modulus of the intermediate material is interpolated as a function of the element density:

$$E(x_i) = E_1 x_i^p \tag{4.2}$$

where E_1 denotes Young's modulus of the solid material and p the penalty exponent. It is assumed that Poisson's ratio is independent of the design variables and the global stiffness matrix $\mathbf{K}$ can be expressed by the elemental stiffness matrix and design variables x_i as

$$\mathbf{K} = \sum_i x_i^p \mathbf{K}_i^0 \tag{4.3}$$

where $\mathbf{K}_i^0$ denotes the elemental stiffness matrix of the solid element.

4.3   Sensitivity Analysis and Sensitivity Number

4.3.1   Sensitivity Analysis

In finite element analysis, the equilibrium equation of a static structure can be expressed as

$$\mathbf{Ku} = \mathbf{f} \tag{4.4}$$

If we consider the mean compliance (defined in Equation (2.3)) as the objective function and assume that the design variable x_i continuously changes from 1 to $x_{\min}$, the sensitivity of the objective function with respect to the change in the design variable is

$$\frac{dC}{dx_i} = \frac{1}{2}\frac{d\mathbf{f}^{\mathrm{T}}}{dx_i}\mathbf{u} + \frac{1}{2}\mathbf{f}^{\mathrm{T}}\frac{d\mathbf{u}}{dx_i} \tag{4.5}$$

The adjoint method will be used to determine the sensitivity of the displacement vector, as explained below. By introducing a vector of Lagrangian multiplier λ, an extra term $\lambda^{\mathrm{T}}(\mathbf{f} - \mathbf{Ku})$ can be added to the objective function without changing anything due to the equilibrium Equation (4.4). Thus

$$C = \frac{1}{2}\mathbf{f}^{\mathrm{T}}\mathbf{u} + \lambda^{\mathrm{T}}(\mathbf{f} - \mathbf{Ku}) \tag{4.6}$$

The sensitivity of the modified objective function can be written as

$$\frac{dC}{dx_i} = \frac{1}{2}\frac{d\mathbf{f}^{\mathrm{T}}}{dx_i}\mathbf{u} + \frac{1}{2}\mathbf{f}^{\mathrm{T}}\frac{d\mathbf{u}}{dx_i} + \frac{d\lambda^{\mathrm{T}}}{dx_i}(\mathbf{f} - \mathbf{Ku}) + \lambda^{\mathrm{T}}\left(\frac{d\mathbf{f}}{dx_i} - \frac{d\mathbf{K}}{dx_i}\mathbf{u} - \mathbf{K}\frac{d\mathbf{u}}{dx_i}\right) \tag{4.7}$$

Note that the third term in the above equation becomes zero due to the equilibrium equation. Also it is assumed that the variation of an element has no effect on the applied load vector and therefore $\frac{d\mathbf{f}}{dx_i} = \mathbf{0}$. Thus, the sensitivity of the objective function becomes

$$\frac{dC}{dx_i} = \left(\frac{1}{2}\mathbf{f}^{\mathrm{T}} - \lambda^{\mathrm{T}}\mathbf{K}\right)\frac{d\mathbf{u}}{dx_i} - \lambda^{\mathrm{T}}\frac{d\mathbf{K}}{dx_i}\mathbf{u} \tag{4.8}$$

From Equation (4.6) it is seen that because $(\mathbf{f} - \mathbf{Ku})$ is equal to zero the Lagrangian multiplier vector λ can be chosen freely. To eliminate the unknown $\frac{d\mathbf{u}}{dx_i}$ from the above sensitivity expression, λ is chosen such that

$$\frac{1}{2}\mathbf{f}^{\mathrm{T}} - \lambda^{\mathrm{T}}\mathbf{K} = \mathbf{0} \tag{4.9}$$

Comparing the above equation to the equilibrium Equation (4.4) reveals that the solution for the Lagrangian multiplier vector λ is

$$\lambda = \frac{1}{2}\mathbf{u} \qquad (4.10)$$

By substituting λ into Equation (4.8), the sensitivity of the objective function becomes

$$\frac{dC}{dx_i} = -\frac{1}{2}\mathbf{u}^T\frac{d\mathbf{K}}{dx_i}\mathbf{u} \qquad (4.11)$$

By substituting the material interpolation scheme Equation (4.3) into the above equation, the sensitivity of the objective function with regard to the change in the ith element can be found as

$$\frac{\partial C}{\partial x_i} = -\frac{1}{2}px_i^{p-1}\mathbf{u}_i^T\mathbf{K}_i^0\mathbf{u}_i \qquad (4.12)$$

4.3.2   Sensitivity Number

In the ESO/BESO methods, a structure is optimized using discrete design variables. That is to say that only two bound materials are allowed in the design. Therefore, the sensitivity number used in the ESO/BESO methods can be defined by the relative ranking of the sensitivity of an individual element as

$$\alpha_i = -\frac{1}{p}\frac{\partial C}{\partial x_i} = \begin{cases} \frac{1}{2}\mathbf{u}_i^T\mathbf{K}_i^0\mathbf{u}_i & \text{when } x_i = 1 \\ \frac{x_{min}^{p-1}}{2}\mathbf{u}_i^T\mathbf{K}_i^0\mathbf{u}_i & \text{when } x_i = x_{min} \end{cases} \qquad (4.13)$$

It is noted that the sensitivity numbers of soft elements depend on the selection of the penalty exponent p. When the penalty exponent tends to infinity, the above sensitivity number becomes

$$\alpha_i == \begin{cases} \frac{1}{2}\mathbf{u}_i^T\mathbf{K}_i^0\mathbf{u}_i & \text{when } x_i = 1 \\ 0 & \text{when } x_i = x_{min} \end{cases} \qquad (4.14)$$

This equation indicates that the sensitivity numbers of solid elements and soft elements are equal to the elemental strain energy and zero, respectively. This is consistent with the sensitivity numbers of the hard-kill BESO method introduced in the previous chapter. According to the material interpolation scheme Equation (4.3), the stiffness $\mathbf{K}$ of soft elements also becomes zero as p approaches infinity. For the above reasons, when p tends to infinity the soft elements are equivalent to void elements and can be completely removed from the design domain as in the hard-kill ESO/BESO methods. Therefore, it is concluded that the hard-kill BESO method is a special case of the soft-kill BESO method where the penalty exponent p approaches infinity. To distinguish from the hard-kill BESO method, the soft-kill BESO method throughout this book will have a finite penalty exponent.

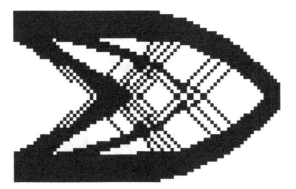

Figure 4.1 'Optimal' design from the soft-kill BESO method without a filter.

The optimality criterion for the problem stated in Equation (4.1) can easily be derived if no restriction is imposed on the design variables x_i, i.e. the strain energy densities of all elements should be equal. Thus, the elements with higher strain energy density should have x_i increased and the elements with lower strain energy density should have x_i decreased. For the soft-kill BESO method, as the design variables are restricted to be either x_{min} or 1, the optimality criterion can be described as that strain energy densities of solid elements are always higher than those of soft elements. Therefore, we devise an update scheme for the design variables x_i by changing from 1 to x_{min} for elements with lowest sensitivity numbers and from x_{min} to 1 for elements with highest sensitivity numbers.

In order to improve the convergence of the soft-kill BESO method, the simple averaging scheme given in Equation (3.8) should be applied to the sensitivity number to take into consideration the historical information of the elemental sensitivity from previous iterations. The averaging scheme will suppress unwarranted changes of design variables for solid elements with high historical sensitivity numbers and void elements with low historical sensitivity numbers.

Different from hard-kill BESO, the soft-kill BESO method can add and recover elements without any help of a filter scheme. Let us reconsider the cantilever example shown in Figure 3.5 using the soft-kill BESO method described above. The design domain is discretized into 80×50 four node plane stress elements. The parameters used are $ER = 1\%$, $x_{min} = 0.001$, $p = 3.0$. The final topology is shown in Figure 4.1. However, due to the checkerboard pattern, the resulting topology has little practical value. Therefore, the BESO filter described in Chapter 3 should be used to suppress the checkerboard pattern and to overcome the mesh-dependency problem.

4.4 Examples

4.4.1 Topology Optimization of a Short Cantilever

Using the same finite element mesh, the above example is solved by the soft-kill BESO method with the filter scheme described in Section 3.3.2. The BESO parameters are: $ER = 2\%$,

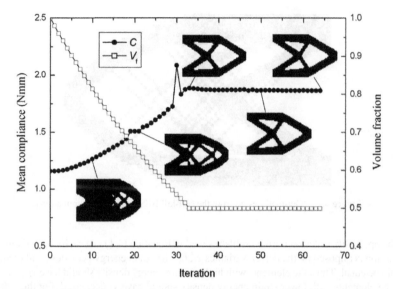

Figure 4.2 Evolutionary histories of mean compliance, volume fraction and topology when $p = 3.0$.

$AR_{\max} = 50\%$, $r_{\min} = 3$ mm and $\tau = 0.1\%$. The lower bound of the material density, $x_{\min}$, is set to be 0.001 which represents the void element in the following simulations.

Figure 4.2 shows the evolutionary histories of the mean compliance, the volume fraction and the topology for the soft-kill BESO method with the penalty exponent $p = 3.0$. It is seen that the mean compliance increases initially as the volume fraction decreases. It then converges to an almost constant value after the objective volume (50%) is achieved. The final design is obtained after 65 iterations.

Figures 4.3(a) and (b) show the BESO optimal designs with $p = 1.5$ and $p = 3.0$ respectively. The mean compliances of both designs are equal to 1.865 Nmm. Compared to the solution obtained from the hard-kill BESO method (the topology in Figure 3.7(f) with a mean compliance of 1.87 Nmm), the results of topologies and mean compliances are remarkably

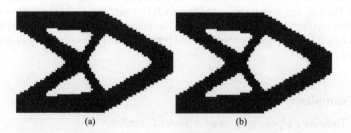

Figure 4.3 Optimal designs from soft-kill BESO method with different penalty exponents: (a) $p = 1.5$; (b) $p = 3.0$.

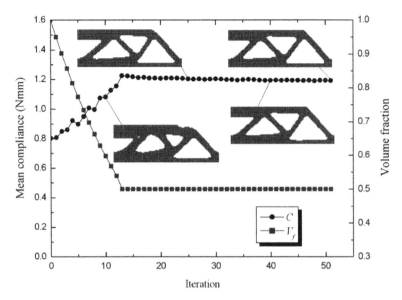

Figure 4.4 Evolution histories of mean compliance, volume fraction and topology when $p = 2.0$.

close. It is noted that the penalty exponent p has almost no effect on the optimal solutions to this problem when it increases from 1.5 to infinity. Usually, however, p must be greater than 1.

4.4.2 Topology Optimization of a Beam

Consider the topology optimization problem of the beam structure shown in Figure 3.9. The symmetric right half of the domain (with 1 mm thickness) is discretized into 120×40 four node plane stress elements. The soft-kill BESO method with the following parameters are used: $ER = 5\%, AR_{\max} = 5\%, r_{\min} = 6$ mm, $x_{\min} = 0.001$ and $\tau = 0.01\%$. Figure 4.4 shows the evolution histories of the mean compliance, the volume fraction and the topology when the penalty exponent $p = 2.0$ is used. The solution converges after 51 iterations.

Figures 4.5(a) and (b) show the final topologies for the soft-kill BESO method with $p = 2.0$ and $p = 10.0$ respectively. The mean compliances of both designs are 1.19 Nmm. The topologies are very similar to the design obtained from the hard-kill BESO method shown in Figure 3.11(f), and the mean compliances are very close to that of the hard-kill solution which is 1.17 Nmm. This example also demonstrates that the penalty exponent p has almost no effect on the optimal solutions once it is sufficiently large.

4.4.3 Topology Optimization of a 3D Cantilever

In this example, we minimize the mean compliance of the 3D cantilever shown in Figure 3.17 subject to a given volume constraint. Due to symmetry, only half of the structure is modelled

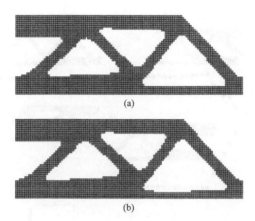

(a)

(b)

Figure 4.5   Optimal designs from soft-kill BESO method with different penalty exponents: (a) $p = 2.0$; (b) $p = 10.0$.

using a total of 40 000 eight node brick elements. The objective volume is 10 % of the total design domain volume. The soft-kill BESO method is used with the following parameters: $ER = 3\%$, $AR_{max} = 50\%$, $r_{min} = 3$ mm, $p = 3.0$, $x_{min} = 0.001$ and $\tau = 0.1\%$. The optimal design shown in Figure 4.6 is obtained after 85 iterations. The mean compliance of the final topology is 1866 Nmm. Both the final topology and the corresponding mean compliance are very close to those of the hard-kill BESO solution shown in Figure 3.19(f). However, the whole computation takes 7 hours and 10 minutes, compared to 2 hours and 40 minutes by the hard-kill BESO method. The huge difference is due to the fact that all soft elements have to

Figure 4.6   Optimal design for 3D cantilever using soft-kill BESO method with $p = 3.0$.

be included in the finite element analyses, unlike the real void elements in the hard-kill BESO which can be ignored.

4.5 Conclusion

This chapter has presented a generalized BESO method utilizing a material interpolation scheme with penalization as per Equation (4.2). To obtain a solid-void design, two bound material densities (1 and x_{min}) are used as discrete design variables. The sensitivity numbers given in Equation (4.13) are derived from the sensitivity analysis. Equations (4.2) and (4.13) indicate that the hard-kill BESO method described in Chapter 3 is a special case of the generalized BESO method where the penalty exponent p tends to infinity.

The capability of the soft-kill BESO method for topology optimization has been demonstrated by several examples. It is seen from the results that the optimal designs from the generalized BESO method are independent of the selection of the penalty exponent (normally $p \geq 1.5$), even when p tends to infinity. In other words, the optimal solutions from the soft-kill BESO method are the same as those from the hard-kill BESO method. Due to its high computational efficiency, the hard-kill BESO method is recommended for stiffness optimization (minimizing mean compliance) problems, rather than the soft-kill BESO method.

It is worth emphasizing that the soft-kill BESO method does not used continuous design variables. The design variables in the soft-kill BESO method can either be 1 or x_{min}. Since x_{min} is usually very small (e.g. 0.001), the soft elements are structurally negligible and therefore the topologies obtained are virtually solid-void designs with no intermediate material densities between 1 and x_{min}. This is in contrast to the SIMP method with continuous design variables which may result in fuzzy topologies with 'grey' elements, as shown in Figures 3.8 and 3.12 ($p = 3$ is used in SIMP for both cases). However, various researchers have shown that the SIMP method with continuous design variables may also produce a solid-void design if the chosen p is sufficiently large (Bendsøe and Sigmund 2003; Sigmund and Petersson 1998; Zhou and Rozvany 1991).

It is noted that when $p = 1$ the optimization problem corresponds to a variable thickness sheet problem rather than a topology optimization problem. In such a case, it is impossible to obtain a convergent solid-void design using the BESO method with discrete design variables because the solution does not actually exist. Numerical tests, including examples shown in this chapter, indicate that $p \geq 1.5$ should be used for the generalized BESO method.

Appendix

This appendix contains a soft-kill BESO MATLAB® code that can be used to solve simple 2D stiffness optimization problems, such as those shown in this chapter. The code is developed based on the 99 line SIMP code (Sigmund 2001). The design domain is assumed to be rectangular and discretized using four node plane stress elements. Here, a short cantilever is taken as an example. Other structures with different loading and boundary conditions can be solved by modifying lines 80–84 of the code. The input data are:

nelx – total number of elements in the horizontal direction;
nely – total number of elements in the vertical direction;

volfrac – volume fraction which defines the ratio of the final volume and the design
   domain volume;
er – evolutionary volume ratio, normally 0.02;
rmin – filter radius, normally 3 (or the size of several elements).

This code (and other BESO software packages) can be downloaded from the website www.
isg.rmit.edu.au, or obtained from the authors by emailing huang.xiaodong@rmit.edu.au or
mike.xie@rmit.edu.au.

```
1 %%%% A SOFT-KILL BESO CODE BY X. HUANG and Y.M. XIE %%%%
2 function sbeso(nelx,nely,volfrac,er,rmin);
3 % INITIALIZE
4 x(1:nely,1:nelx) = 1.; vol=1.; i = 0; change = 1.; penal = 3.;
5 % START iTH ITERATION
6 while change > 0.001
7 i = i + 1; vol = max(vol*(1-er),volfrac);
8 if i >1; olddc = dc; end
9 % FE-ANALYSIS
10 [U]=FE(nelx,nely,x,penal);
11 % OBJECTIVE FUNCTION AND SENSITIVITY ANALYSIS
12 [KE] = lk;
13 c(i) = 0.;
14 for ely = 1:nely
15  for elx = 1:nelx
16   n1 = (nely+1)*(elx-1)+ely;
17   n2 = (nely+1)* elx +ely;
18   Ue = U([2*n1-1;2*n1; 2*n2-1;2*n2; 2*n2+1;2*n2+2;
2*n1+1;2*n1+2],1);
19   c(i) = c(i) + 0.5*x(ely,elx)^penal*Ue'*KE*Ue;
20   dc(ely,elx) = 0.5*x(ely,elx)^(penal-1)*Ue'*KE*Ue;
21  end
22 end
23 % FILTERING OF SENSITIVITIES
24 [dc] = check(nelx,nely,rmin,x,dc);
25 % STABLIZATION OF EVOLUTIONARY PROCESS
26 if i > 1; dc = (dc+olddc)/2.; end
27 % BESO DESIGN UPDATE
28 [x]  = ADDDEL(nelx,nely,vol,dc,x);
29 % PRINT RESULTS
30 if i>10;
31 change=abs(sum(c(i-9:i-5))-sum(c(i-4:i)))/sum(c(i-4:i));
32 end
33 disp([' It.: ' sprintf('%4i',i) ' Obj.: '
sprintf('%10.4f',c(i)) ...
34    ' Vol.: ' sprintf('%6.3f',sum(sum(x))/(nelx*nely)) ...
35    ' ch.: ' sprintf('%6.3f',change )])
36 % PLOT DENSITIES
37 colormap(gray); imagesc(-x); axis equal; axis tight;
axis off;pause(1e-6);
```

```
38 end
39 %%%%%%%%% OPTIMALITY CRITERIA UPDATE
%%%%%%%%%%%%%%%%%%%%%%%%%%%%%%%%%%%%%%%%%%%
40 function [x]=ADDDEL(nelx,nely,volfra,dc,x)
41 l1 = min(min(dc)); l2 = max(max(dc));
42 while ((l2-l1)/l2 > 1.0e-5)
43  th = (l1+l2)/2.0;
44  x = max(0.001,sign(dc-th));
45  if sum(sum(x))-volfra*(nelx*nely) > 0;
46   l1 = th;
47  else
48   l2 = th;
49  end
50 end
51 %%%%%%%%% MESH-INDEPENDENCY FILTER
%%%%%%%%%%%%%%%%%%%%%%%%%%%%%%%%%%%%%%%%%%%
52 function [dcf]=check(nelx,nely,rmin,x,dc)
53 dcf=zeros(nely,nelx);
54 for i = 1:nelx
55 for j = 1:nely
56  sum=0.0;
57  for k = max(i-floor(rmin),1):min(i+floor(rmin),nelx)
58   for l = max(j-floor(rmin),1):min(j+floor(rmin),nely)
59    fac = rmin-sqrt((i-k)^2+(j-1)^2);
60    sum = sum+max(0,fac);
61    dcf(j,i) = dcf(j,i) + max(0,fac)*dc(l,k);
62   end
63  end
64  dcf(j,i) = dcf(j,i)/sum;
65 end
66 end
67 %%%%%%%%% FE-ANALYSIS
%%%%%%%%%%%%%%%%%%%%%%%%%%%%%%%%%%%%%%%%%%%%%%%%%%%
68 function [U]=FE(nelx,nely,x,penal)
69 [KE] = lk;
70 K = sparse(2*(nelx+1)*(nely+1), 2*(nelx+1)*(nely+1));
71 F = sparse(2*(nely+1)*(nelx+1),1); U =
zeros(2*(nely+1)*(nelx+1),1);
72 for elx = 1:nelx
73 for ely = 1:nely
74  n1 = (nely+1)*(elx-1)+ely;
75  n2 = (nely+1)* elx +ely;
76  edof = [2*n1-1; 2*n1; 2*n2-1; 2*n2; 2*n2+1; 2*n2+2; 2*n1+1;
2*n1+2];
77  K(edof,edof) = K(edof,edof) + x(ely,elx)^penal*KE;
78 end
79 end
80 % DEFINE LOADS AND SUPPORTS (Cantilever)
81 F(2*(nelx+1)*(nely+1)-nely,1)=-1.0;
82 fixeddofs=[1:2*(nely+1)];
```

```
83 alldofs   = [1:2*(nely+1)*(nelx+1)];
84 freedofs  = setdiff(alldofs,fixeddofs);
85 % SOLVING
86 U(freedofs,:) = K(freedofs,freedofs) \ F(freedofs,:);
87 U(fixeddofs,:)= 0;
88 %%%%%%%%%% ELEMENT STIFFNESS MATRIX
%%%%%%%%%%%%%%%%%%%%%%%%%%%%%%%%%%%%%%%%%%%%
89 function [KE]=lk
90 E = 1.;
91 nu = 0.3;
92 k=[ 1/2-nu/6 1/8+nu/8 -1/4-nu/12 -1/8+3*nu/8 ...
93  -1/4+nu/12 -1/8-nu/8 nu/6    1/8-3*nu/8];
94 KE = E/(1-nu^2)*[ k(1) k(2) k(3) k(4) k(5) k(6) k(7) k(8)
95         k(2) k(1) k(8) k(7) k(6) k(5) k(4) k(3)
96         k(3) k(8) k(1) k(6) k(7) k(4) k(5) k(2)
97         k(4) k(7) k(6) k(1) k(8) k(3) k(2) k(5)
98         k(5) k(6) k(7) k(8) k(1) k(2) k(3) k(4)
99         k(6) k(5) k(4) k(3) k(2) k(1) k(8) k(7)
100        k(7) k(4) k(5) k(2) k(3) k(8) k(1) k(6)
101        k(8) k(3) k(2) k(5) k(4) k(7) k(6) k(1)];
```

References

Bendsøe, M.P. (1989). Optimal shape design as a material distribution problem. *Struct. Optim.* **1**: 193–202.

Bendsøe, M.P. and Sigmund, O. (1999). Material interpolation schemes in topology optimization. *Archive Appl. Mech.* **69**: 635–54.

Bendsøe, M.P. and Sigmund, O. (2003). *Topology Optimization: Theory, Method and Application.* Berlin: Springer.

Hinton, E. and Sienz, J. (1995). Fully stressed topological design of structures using an evolutionary procedure. *Eng. Comput.* **12**: 229–44.

Huang, X. and Xie, Y.M. (2009). Bi-directional evolutionary topology optimization of continuum structures with one or multiple materials. *Comput. Mech.* **43**: 393–401.

Rietz, A. (2001). Sufficiency of a finite exponent in SIMP (power law) methods. *Struct. Multidisc. Optim.* **21**: 159–63.

Rozvany, G.I.N. and Querin, O.M. (2002). Combining ESO with rigorous optimality criteria. *Int. J. Vehicle Design* **28**: 294–9.

Sigmund, O. and Petersson, J. (1998). Numerical instabilities in topology optimization: A survey on procedures dealing with checkerboards, mesh-dependencies and local minima. *Struct. Optim.* **16**: 68–75.

Sigmund, O. (2001). A 99 line topology optimization code written in MATLAB. *Struct. Multidisc. Optim.* **21**: 120–7.

Zhou, M. and Rozvany, G.I.N. (1991). The COC algorithm, Part II: topological, geometrical and generalized shape optimization. *Comput. Meth. Appl. Mech. Engng.* **89**: 309–36.

Zhu, J.H., Zhang, W.H. and Qiu, K.P. (2007). Bi-directional evolutionary topology optimization using element replaceable method. *Comput. Mech.* **40**: 97–109.

5

Comparing BESO with Other Topology Optimization Methods

5.1 Introduction

In the past few decades, significant progress has been made in the theory and application of topology optimization. Apart from ESO/BESO, other representative methods include the homogenization method (Bendsøe and Kikuchi 1988; Bendsøe and Sigmund 2003), the solid isotropic material with penalization (SIMP) method (Bendsøe 1989; Zhou and Rozvany 1991; Sigmund 1997; Rietz 2001; Bendsøe and Sigmund 2003), and the level set method (Sethian and Wiegmann 2000; Wang *et al.* 2003; Allaire *et al.* 2004; Wang *et al.* 2004).

The homogenization method introduces micro-perforated composites as admissible designs to relax the originally ill-posed 0/1 (void/solid) optimization problem (Bendsøe and Sigmund 2003). A relationship between microstructure parameters and effective material properties is established by the homogenization theory. The optimal material distribution may be obtained by determining the size parameters of the microstructure according to an optimality criteria procedure. However, the resulting structure cannot be built directly since no definite length-scale is associated with the microstructure. This approach to topology optimization is of importance to the extent that the solution provides mathematical bounds to the theoretical performance of a structure.

An alternative approach is the SIMP method which is based on the assumption that each element contains an isotropic material with variable density as described in Chapter 4. The elements are used to discretize the design domain and the design variables are the relative densities of the elements. Then, the power-law interpolation scheme (4.2) penalizes the intermediate densities to obtain a solution with nearly 0/1 material distribution. Here, the SIMP method is defined by its underlying principle rather than the particular interpolation scheme.

The level set method is another well-established topology optimization approach. It is basically a steepest descent method by combining the shape sensitivity analysis with the Hamilton-Jacobi equation for moving the level-set function, for doing topology design of structures (Wang *et al.* 2003; 2004).

Evolutionary Topology Optimization of Continuum Structures: Methods and Applications   Xiaodong Huang and Mike Xie
© 2010 John Wiley & Sons, Ltd

Currently, the SIMP method has demonstrated its effectiveness in a broad range of examples and its algorithm has been widely accepted due to its computational efficiency and conceptual simplicity (Sigmund 2001). In the following sections, a comparison between the SIMP method and the ESO/BESO methods will be conducted.

5.2   The SIMP Method

To minimize the structural compliance using the SIMP method, the topology optimization problem can be written as

$$\text{Minimize } c = \mathbf{f}^{\mathbf{T}}\mathbf{u} \tag{5.1a}$$

$$\text{Subject to}: V^* - \sum_{i=1}^{N} V_i x_i = 0 \tag{5.1b}$$

$$0 < x_{\min} \leq x_i \leq 1 \tag{5.1c}$$

where the compliance, c, is twice as much as the mean compliance C defined in Equation (3.1) for BESO. Here, the design variable x_i varies continuously from $x_{\min}$ to 1 instead of being either of the two discrete bound values as used in the BESO method. Based on the material interpolation scheme (4.2), the global stiffness matrix can be expressed as

$$\mathbf{K} = \sum_i x_i^p \mathbf{K}_i^0 \tag{5.2}$$

To obtain a clear 0/1 design, normally $p \geq 3$ is used in the SIMP method.

Following the derivation in Chapter 4, the sensitivity of the objective function c is found to be

$$\frac{\partial c}{\partial x_i} = -p x_i^{p-1} \mathbf{u}_i^T \mathbf{K}_i^0 \mathbf{u}_i \tag{5.3}$$

The above optimization problem can be solved using several different approaches such as the Optimality Criteria (OC) methods (Zhou and Rozvany 1991; Rozvany and Zhou 1994), the Method of Moving Asymptotes (MMA) (Svanberg 1987) and some others. A standard OC updating scheme for the design variables can be formulated as (Bendsøe and Sigmund 2003)

$$x_i^{K+1} = \begin{cases} \max(x_{\min}, \ x_i^K - m) & \text{if } x_i^K B_i^\eta \leq \max(x_{\min}, \ x_i^K - m) \\ \min(1, \ x_i^K + m) & \text{if } \min(1, \ x_i^K + m) \leq x_i^K B_i^\eta \\ x_i^K B_i^\eta & \text{otherwise} \end{cases} \tag{5.4}$$

where x_i^K denotes the value of the design variable at iteration K, m is the positive move limit, η is a numerical damping coefficient (typically equal to 0.5) and B_i is found from the optimality condition as

$$B_i = \lambda^{-1} p x_i^{p-1} \mathbf{u}_i^T \mathbf{K}_i^0 \mathbf{u}_i \tag{5.5}$$

where λ is a Lagrangian multiplier that can be determined using a bisection method or a Newton method (Bendsøe and Sigmund 2003).

To ensure that the optimal design is mesh-independent and checkerboard free, Sigmund (1997; 2001) introduces the following filter scheme by modifying elemental sensitivities:

$$\frac{\partial c}{\partial x_i} = \frac{1}{x_i \sum\limits_{j=1}^{N} H_{ij}} \sum_{j=1}^{N} H_{ij} x_j \frac{\partial c}{\partial x_j} \qquad (5.6)$$

where N is the total number of elements in the mesh and H_{ij} is the mesh-independent weight factor defined as

$$H_{ij} = r_{min} - r_{ij}, \quad \{i \in N \,|\, r_{ij} \leq r_{min}\}$$

where r_{ij} is the distance between the centres of elements i and j. The weight factor H_{ij} is zero outside the circular filter area shown in Figure 3.2.

With the introduction of material penalization in Equation (5.2), the original optimization problem (5.1) becomes nonconvex, and therefore, it is possible to result in a local optimum with grey regions. Numerical examples show that a local optimum may be avoided by applying the *continuation method* (Rozvany *et al.* 1994; Sigmund and Petersson 1998). Theoretically, a global optimum cannot be guaranteed even using the continuation method as noted by Stolpe and Svanberg (2001). Several continuation procedures have been proposed by various researchers.

Rozvany *et al.* (1994) and Rozvany (2009) suggest a continuation method by using an unpenalized material ($p = 1$) in the first computational cycle, and then increasing the penalty exponent progressively in small steps in subsequent cycles. In this way, a global optimum with $p = 1$ is derived for the originally convex problem, and later, grey regions change locally into black-and-white areas of the same average density; and thereby, the solution is not moved too far away from the global optimum. A different continuation method using a mesh-independency filter is suggested by Sigmund (1997) and Sigmund and Torquato (1997) where the computation starts with a large value of the filter radius r_{min} to ensure a convex solution at the beginning and then to decrease r_{min} gradually until the solution ends up with a 0/1 design. In this chapter, both approaches of the continuation method as described above will be tested.

5.3 Comparing BESO with SIMP

The filter scheme is a heuristic technique for overcoming the checkerboard and mesh-dependency problems in topology optimization. It should be noted that the filtering process averages and smoothes out the otherwise discontinuous elemental sensitivity. Therefore, it is better to compare BESO with SIMP algorithms at two levels – one without the mesh-independency filter and the other with it.

A long cantilever as shown in Figure 5.1 is selected as a test example because it involves a series of bars broken during the evolution process of ESO/BESO. A concentrated load $F = 1$N is applied downwards in the middle of the free end. Young's modulus $E = 1$MPa and Poisson's ratio $v = 0.3$ are assumed. The design domain is discretized into 160×40 four node plane stress elements.

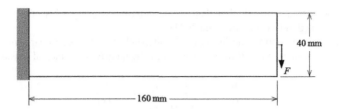

Figure 5.1   Design domain of a long cantilever.

5.3.1   Comparison of Topology Optimization Algorithms without a Mesh-independency Filter

The used parameters and final solutions for various topology optimization methods are listed in Table 5.1. Without using a filter, the topologies obtained from these methods are quite different. In fact they are hardly comparable. It is difficult to tell which topology is the best unless the value of the final objective function is compared. (Here, the mean compliance is used as the objective function for all the topology optimization methods.) It is observed that the continuation method produces the lowest mean compliance among all the optimization methods although it takes the largest number of iterations. ESO and soft-kill BESO require much less iterations and result in mean compliances that are close to that of the continuation method. Note that hard-kill BESO without the filter degenerates to ESO. The final mean compliance from the SIMP method with $p = 3$ is much higher than that of other methods. However, this does not necessarily mean that SIMP is worse than other methods, because the higher mean compliance may have been caused by the over-estimated strain energy of elements with intermediate densities in the SIMP topology.

It should be noted that the ESO method usually requires a fine mesh, especially for a problem with a low final volume fraction. Normally, a smaller *ER* results in a better solution. The computational efficiency of ESO highly depends on the selected parameters such as evolutionary ratio *ER* and the mesh size. In most cases, the ESO method using a small *ER* and a fine mesh can produce a good solution. This is a merit of the original ESO procedure.

Compared to ESO, soft-kill BESO and SIMP methods are more stable and less dependent on the selected parameters although a relatively fine mesh is still required by the soft-kill BESO method. Provided that the penalty exponent *p* is large enough, both soft-kill BESO and SIMP methods can produce good results in most cases since the final solutions from these two methods meet the respective optimality criteria.

5.3.2   Comparison of Topology Optimization Algorithms with a Mesh-independency Filter

The above problem is reanalysed using the four topology optimization methods but this time with a mesh-independency filter. The filter schemes (3.6) and (5.6) are used for BESO and SIMP, respectively. It is noted that these two filter schemes are quite similar. Table 5.2 lists the used parameters and the solutions obtained. It is seen that the four optimization algorithms produce very similar topologies except that the SIMP design has some grey areas of intermediate material densities. For practical applications, the topologies in Table 5.2 with a clear definition of each member are far more useful than those shown in Table 5.1.

Table 5.1 Comparison of topology optimization methods without a mesh-independency filter.

	Optimization parameters	Total iteration	Solutions (for volume fraction of 50 %)		Error[1] (%)
ESO[2]	$ER = 1\%$	67		$C = 188.91$ Nmm	4.12
Soft-kill BESO	$ER = 2\%$ $p = 3.0$	44		$C = 183.25$ Nmm	1.00
SIMP	$m = 0.02$ $p = 3.0$	37		$C = 196.48$ Nmm	8.29
Continuation	$p_{initial} = 1$ $p = 0.1$ $p_{end} = 5.0$	337		$C = 181.44$ Nmm	–

[1] This refers to the error of the mean compliance C as compared to the result of the continuation method.
[2] Hard-kill BESO without a mesh-independency filter degenerates to ESO.

Table 5.2 Comparison of topology optimization methods with a mesh-independency filter.

	Optimization parameters	Total iteration	Solutions (for volume fraction of 50 %)		Error[1] (%)
Hard-kill BESO[2]	$ER = 2\%$ $AR_{max} = 50\%$ $r_{min} = 3.0\,\text{mm}$	52		$C = 181.79$ Nmm	0.61
Soft-kill BESO	$ER = 2\%$ $p = 3.0$ $r_{min} = 3.0\,\text{mm}$	46		$C = 181.71$ Nmm	0.56
SIMP	$m = 0.02$ $p = 3.0$ $r_{min} = 3.0\,\text{mm}$	44		$C = 201.70$ Nmm	11.63
Continuation	$r_{min}^{ini} = 3.0\,\text{mm}$ $\Delta r_{min} = 0.1\,\text{mm}$ $r_{min}^{end} = 1.0\,\text{mm}$	267		$C = 180.69$ Nmm	–

[1] This refers to the error of the mean compliance C as compared to the result of the continuation method.
[2] The penalty tends to infinity for hard-kill BESO method.

The mean compliances of both hard-kill and soft-kill BESO solutions are very close to that of the continuation method. However, the continuation method requires more than five times as many iterations as the BESO methods. Although hark-kill BESO takes slightly more iterations than soft-kill BESO, the former algorithm is actually the quickest because the hard-killed elements are not included in the finite element analysis.

Again, the SIMP method with $p = 3$ converges to a solution with the highest mean compliance among all methods, for the same reason as we have explained in the previous section.

Figure 5.2 shows the evolution histories of the objective function for the four topology optimization methods. The mean compliances of both hard-kill and soft-kill BESO methods increases, with occasionally abrupt jumps (due to breaking up of some bars), as the total volume gradually decreases. After about 35 iterations the volume fraction reaches its target of 50 %. In subsequent iterations, while the volume remains unchanged the mean compliance gradually converges to a constant value. Different from the BESO methods, both SIMP and continuation methods have the volume constraint satisfied all the time. While the volume is kept constant from the very beginning, the mean compliances in SIMP and continuation methods decrease gradually until a convergence criterion is satisfied.

In this example and the one in the previous section, the convergence criterion (3.13) is used for the hard-kill and soft-kill BESO methods, with $\tau = 0.001$ and $N = 5$. ESO does not use a convergence criterion – once the final volume fraction is reached the solution process stops. In the SIMP and continuation methods, a different convergence criterion is used where the change in the design variables (instead of the objective function) is monitored. Therefore, strictly speaking, the iteration numbers of the ESO/BESO methods and those of the SIMP and continuation methods shown in Tables 5.1 and 5.2 are not actually comparable. These iteration numbers are only indicative of the computational efficiencies of different optimization methods. Nonetheless, it is safe to say that the continuation method usually takes far more iterations and produces better solutions than other methods considered in this chapter.

5.3.3   Advantages of the BESO Method and Questions yet to be Resolved

The above example shows that the BESO method with a mesh-independency filter is capable of producing high quality designs which are very close to those of the continuation method but with far less iterations. Usually the total number of iterations is less than 100. The current BESO method for stiffness optimization has the following advantages:

- High quality topology solutions
- Excellent computational efficiency
- Algorithms easy to understand and simple to implement

Examples in this and previous chapters show that the hard-kill and soft-kill BESO methods produce almost identical solutions. Compared to the soft-kill BESO method, the hard-kill BESO method is computationally far more efficient because the hard-killed elements are not involved in the finite element analysis. Also, soft elements in the soft-kill BESO method may cause the global stiffness matrix to become ill-conditioned, particularly for a nonlinear structure (Bruns and Tortorelli 2003). For these reasons, the hard-kill BESO method is preferable, especially for complex 3D structures.

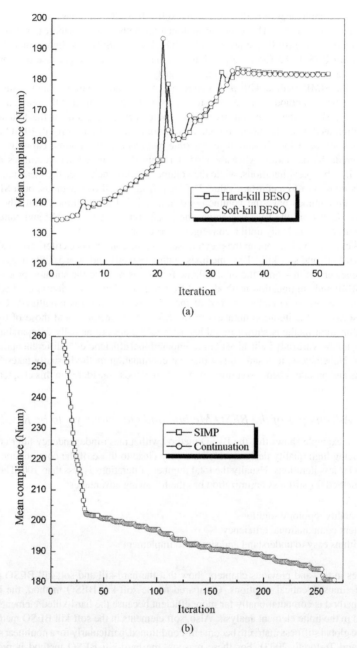

Figure 5.2 Evolution histories of mean compliance: (a) hard-kill and soft-kill BESO methods; (b) SIMP and continuation methods.

The BESO algorithm has been successfully applied to topology optimization problems with a single volume constraint. However, there are still many unresolved questions about the efficacy of BESO due to its simple updating scheme for the design variables and also the discrete nature of the design variables. For example, will it be feasible to apply BESO to problems with a different constraint or multiple constraints (Sigmund 2001; Tanskanen 2002; Rozvany 2009)? To answer some of these questions we shall present extended BESO algorithms for various other topology optimization problems in the next chapter. In the meantime, we shall address another notable question about ESO/BESO which has arisen following the work of Zhou and Rozvany (2001) in which a highly inefficient solution to a cantilever tie-beam structure is presented.

5.4 Discussion on Zhou and Rozvany (2001) Example

5.4.1 Introduction of Zhou and Rozvany (2001) Example

The structure shown in Figure 5.3(a) is used by Zhou and Rozvany (2001) to show the breakdown of hard-kill optimization methods, such as ESO/BESO. In the example, Young's modulus is taken as unity and Poisson's ratio as zero. If hard-kill ESO/BESO methods are applied to an FE model with 100 four node plane stress elements shown in Figure 5.3(b), the element in the vertical tie with the lowest strain energy density will be removed from the ground structure. The mean compliance of the resulting structure will be much higher than that of any institutive design obtained by removing one element in the horizontal beam.

After removing an element in the vertical tie, the resultant structure becomes a cantilever where the vertical load is transmitted by flexural action. The region with the highest strain energy density is at the left-bottom of the cantilever. According to the BESO algorithm, an element may be added in that region rather than recovering the removed element in the vertical tie.

Therefore, Zhou and Rozvany (2001) conclude that hard-kill optimization methods such as ESO/BESO may produce a highly nonoptimal solution. In fact, soft-kill optimization algorithms such as the level set method using continuous design variables may also produce a similar result (Norato et al. 2007). To overcome this problem, the essence of such a solution needs to be examined first.

5.4.2 Is it a Nonoptimal or a Local Optimal Solution?

Obviously, the answer cannot be easily found by simply comparing the values of the objective function. Let us reconsider the above example with a volume fraction of 96%. Hard-kill optimization methods such as ESO will remove the four elements from the vertical tie as shown in Figure 5.3(c). This design is certainly far less efficient than an intuitive design which removes four elements from the horizontal beam.

It is known that the SIMP method with continuous design variables guarantees that its solution should be at least a local optimum. Therefore, this topology optimization problem is tested by the SIMP method starting from an initial guess design (with $x_i = 1$ for all elements in the horizontal beam and $x_i = x_{min} = 0.001$ for the four elements in the vertical tie). It is found that when $p \geq 3.1$ the final solution converges to the structure shown in Figure 5.3(d), which is exactly the same as the initial guess design. Because x_{min} is small, the SIMP solution in Figure 5.3(d) can be considered to be identical to the ESO/BESO solution in Figure 5.3(c).

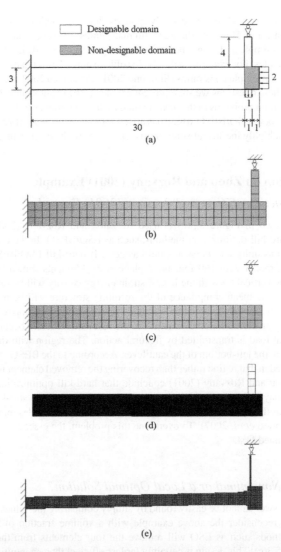

Figure 5.3 Zhou and Rozvany (2001) example: (a) design domain, load and support conditions; (b) a coarse mesh; (c) a highly inefficient local optimum for $V_f = 96\%$ from ESO; (d) a highly inefficient local optimum for $V_f = 0.96$ from SIMP when $p \geq 3.1$; (e) an optimal solution for $V_f = 50\%$ from BESO (Huang and Xie 2008).

These results demonstrate that the above solutions from ESO/BESO and SIMP are essentially a local optimum rather than a nonoptimum. Theoretically it may be more appropriate to call such a solution *a highly inefficient local optimum* than a nonoptimum.

The occurrence of the above 0/1 local optimal design is caused by the large penalty p in the optimization algorithms. Hard-kill ESO/BESO methods have an equivalent penalty of infinity and therefore fail to obtain a better solution once they reach the highly inefficient local optimum. Similarly, the soft-kill BESO method with a finite penalty may also fail because a large penalty ($p \geq 1.5$) is normally required for topology optimization.

The exact value of the penalty p that is large enough to cause a local optimum is dependent upon the optimization problem. For the original Zhou and Rozvany (2001) example given in Figure 5.3(a), the SIMP method will produce a much more efficient solution than the one shown Figure 5.3(d) when $p = 3$ is used. However, if we modify the original problem slightly by reducing the vertical load from 1 to 0.5, the SIMP method with $p = 3$ will again result in the highly inefficient local optimum shown in Figure 5.3(d).

5.4.3 Avoidance of Highly Inefficient Local Optimum

It is well-known that most topology optimization problems are not convex and may have several different local optima. At the same time, most global optimization methods seem to be unable to handle problems of the size of a typical topology optimization problem (Bendsøe and Sigmund 2003). Therefore it is unwarranted to completely dismiss the merit of any optimization method just because it may sometimes produce a local optimum.

Once the essence of the problem is identified (i.e. the optimization algorithm falls into a local optimum), the problem can be avoided by suppressing the local optimum outside the optimization algorithm. For example, Huang and Xie (2008, 2009) suggest checking the boundary and the loading conditions after each design iteration. Once the predefined boundary or loading condition is changed due to element elimination, the optimization program should not proceed any further until some corrective measures are put in place. One effective measure is to use a much finer mesh to discretize the initial design.

For the Zhou and Rozvany example (2001) with a 50 % volume fraction, an optimal solution shown in Figure 5.3(e) has been obtained by the hard-kill BESO method using a very fine mesh (Huang and Xie 2008). It has a mean compliance of 378.4, which is close to that of a simple beam-tie solution proposed by Zhou and Rozvany (2001) of which the mean compliance is 387.5. Huang and Xie (2008) also obtain an optimal solution using the ESO method with the same fine mesh. The ESO solution has a mean compliance of 380.5, which is slightly worse than the BESO result. It is noted that the technique of mesh refinement has been used to tackle this type of problem by several other researchers too (Edwards *et al.* 2007; Norato *et al.* 2007; Zhu *et al.* 2007; Liu *et al.* 2008).

5.5 Conclusion

This chapter has compared various aspects of the current ESO/BESO methods with those of the SIMP method and the continuation method.

Without a mesh-independency filter, the ESO method with an infinite penalty may produce a good solution when the optimization parameters are properly chosen. However, the ESO solution is highly dependent on the used parameters and may sometimes become unstable. In contrast, the BESO method with a finite penalty usually yields a more stable and better solution with an objective function very close to that of the continuation method.

With a mesh-independency filter, all four topology optimization methods considered in this chapter produce almost identical topologies. The values of the objective function from both soft-kill and hard-kill BESO algorithms are much lower than that of the SIMP method and very close to that of the continuation method. The main advantage of the present BESO method is that less iterations are required to obtain high quality designs. The hard-kill BESO method produces an almost identical solution to that of the soft-kill BESO method. The effect of the penalty exponent p becomes negligible once it exceeds a certain value. Therefore, in view of its high computational efficiency, the use of the hard-kill BESO method is highly recommended for the compliance minimization problems.

Like many other optimization methods, BESO cannot guarantee to obtain a global optimum. To avoid a highly inefficient local optimum, it is suggested to check the boundary and the loading conditions in each design iteration. Generally, if a breakdown of boundary support is detected during the optimization process, it may well indicate that the used mesh is too coarse and a finer mesh should be adopted for the initial design in order to achieve a satisfactory solution.

References

Allaire, G., Jouve, F. and Toader, A. M. (2004). Structural optimization using sensitivity analysis and a level-set method. *J. Comput. Phys.* **194**: 363–93.

Bendsøe, M.P. (1989). Optimal shape design as a material distribution problem. *Struct. Optim.* **1**: 193–202.

Bendsøe, M.P. and Kikuchi, N. (1988). Generating optimal topologies in structural design using a homogenization method. *Comput. Meth. Appl. Mech. Engng.* **71**: 197–224.

Bendsøe, M.P. and Sigmund, O. (2003). *Topology Optimization: Theory, Method and Application.* Berlin: Springer.

Bruns, T.E. and Tortorelli, D.A. (2003). An element removal and reintroduction strategy for the topology optimization of structures and compliant mechanisms. *Int. J. Numer. Meth. Engng.* **57**: 1413–30.

Edwards, C.S., Kim, H.A. and Budd, C.J. (2007). An evaluative study on ESO and SIMP for optimising a cantilever tie-beam. *Struct. Multidisc. Optim.* **34**: 403–14.

Huang, X. and Xie, Y.M. (2008). A new look at ESO and BESO optimization methods. *Struct. Multidisc. Optim.* **35**: 89–92.

Huang, X. and Xie, Y.M. (2009). A further review of ESO type methods for topology optimization. *Struct. Multidisc. Optim.* (to appear).

Liu, X., Yi, W.J., Li, Q.S. and Shen, P.S. (2008). Genetic evolutionary structural optimization. *J. Constructional Steel Research* **64**: 305–11.

Norato, J.A., Bendsøe, M.P., Harber, R.B. and Tortorelli, D.A. (2007). A topological derivative method for topology optimization. *Struct. Multidisc. Optim.* **33**: 375–86.

Rietz, A. (2001). Sufficiency of a finite exponent in SIMP (power law) methods. *Struct. Multidisc. Optim.* **21**: 159–63.

Rozvany, G.I.N. (2009). A critical review of established methods of structural topology optimization. *Struct. Multidisc. Optim.* **37**(3): 217–37.

Rozvany, G.I.N. and Zhou, M. (1994). Optimality criteria methods for large structural systems. In: H. Adeli (ed.) *Advances in Design Optimization.* London: Chapman & Hall, pp. 41–108.

Rozvany, G.I.N., Zhou, M. and Sigmund, O. (1994). Optimization of topology. *Advances in Design Optimization.* H. Adeli (ed.), Chapman & Hall, London: 340–99.

Sethian, J.A. and Wiegmann, A. (2000). Structural boundary design via level set and immersed interface methods. *J. Comput. Phys.* **163**: 489–528.

Sigmund, O. (1997). On the design of compliant mechanisms using topology optimization. *Mech. Struct. Mach.* **25**: 495–526.

Sigmund, O. (2001). A 99 line topology optimization code written in MATLAB®. *Struct. Multidisc. Optim.* **21**: 120–7.

Sigmund, O. and Petersson, J. (1998). Numerical instabilities in topology optimization: A survey on procedures dealing with checkerboards, mesh-dependencies and local minima. *Struct. Optim.* **16**: 68–75.

Sigmund, O. and Torquato, S. (1997). Design of materials with extreme thermal expansion using a three-phase topology optimization method. *J. Mech. Phys. Solids* **45**: 1037–67.

Stolpe, M. and Svanberg, K. (2001). On the trajectories of penalization methods for topology optimization. *Struct. Multidisc. Optim.* **21**: 128–39.

Svanberg, K. (1987). The method of moving asymptotes – a new method for structural optimization. *Int. J. Numer. Meth. Engrg.* **24**: 359–73.

Tanskanen, P. (2002). The evolutionary structural optimization method: theoretical aspects. *Comput. Meth. Appl. Mech. Engng.* **191**: 5485–98.

Wang, M. Y., Wang, X. and Guo, D. (2003). A level set method for structural topology optimization. *Comput. Meth. Appl. Mech. Engrg.* **192**: 227–46.

Wang, X., Wang, M. Y. and Guo, D. (2004). Structural shape and topology optimization in a level-set-based framework of region representation. *Struct. Multidisc. Optim.* **27**: 1–19.

Zhou, M. and Rozvany, G.I.N. (1991). The COC algorithm, Part II: topological, geometrical and generalized shape optimization. *Comput. Meth. Appl. Mech. Engng.* **89**: 309–36.

Zhou, M. and Rozvany, G.I.N. (2001). On the validity of ESO type methods in topology optimization. *Struct. Multidisc. Optim.* **21**: 80–3.

Zhu, J.H., Zhang, W.H. and Qiu, K.P. (2007). Bi-directional evolutionary topology optimization using element replaceable method. *Comput. Mech.* **40**: 97–109.

6

BESO for Extended Topology Optimization Problems

6.1   Introduction

In previous chapters we have presented the soft-kill and hard-kill BESO methods for stiffness optimization with a single structural volume constraint. It is noted that the hard-kill BESO method is a special case of the soft-kill BESO method. Various algorithms are incorporated in the BESO methods to overcome numerical problems such as checkerboard pattern, mesh-dependency, and lack of convergence of the solution. These algorithms are of significant importance when we extend BESO to other applications.

Topology optimization problems may sometimes consider objective functions other than the stiffness and constraints other than the structural volume. As discussed in previous chapters, for optimization techniques based on finite element analysis, elements acting as the design variables cannot be directly eliminated from the design domain unless soft elements are fully equivalent to void elements. In other words, for a new topology optimization problem, it may be more reliable to develop a soft-kill BESO method first and then explore the possibility of a corresponding hard-kill approach. Following this route, we will extend the BESO method to various topology optimization problems in this chapter.

Based on the given problem statement on the optimization objective and constraints, the sensitivity of the objective function with respect to the design variables should be determined according to the assumed material interpolation scheme. Then, sensitivity numbers which provide the relative ranking of elemental sensitivities are established for discrete design variables. Following the BESO procedure presented in previous chapters, the discrete design variables are updated according to the sensitivity numbers. As a result, the structure evolves to an optimum automatically.

This chapter includes the recent research of the authors into various topology optimization problems (Huang and Xie 2009a; 2009b; 2009c; 2009d; 2009e; Huang *et al.* 2009). The developed BESO algorithms can be directly applied to many practical design problems, some of which are presented in Chapter 9. In this chapter, however, we illustrate the techniques and possible applications with a series of test examples.

Evolutionary Topology Optimization of Continuum Structures: Methods and Applications   Xiaodong Huang and Mike Xie
© 2010 John Wiley & Sons, Ltd

6.2 Minimizing Structural Volume with a Displacement or Compliance Constraint

In this optimization problem, the objective is to save material in a structure and the constraint is imposed on the mean compliance or the displacement. It has a significant application in practice. For example, in certain designs the maximum deflection of the structure must be less than a certain value. The main challenge is that the BESO method evolves the structure by removing and adding materials. Thus, a volume constraint rather than a mean compliance or displacement constraint can be easily implemented. As observed by Rozvany (2009), the previous BESO procedure cannot be readily used for optimization problems with constraints other than the structural volume.

The topology optimization problem based on finite element analysis may be stated as

$$
\text{Minimize } V = \sum_{i=1}^{N} V_i x_i
$$
$$
\text{Subject to: } u_j = u_j^*
$$
$$
x_i = x_{\min} \text{ or } 1
$$

(6.1)

where u_j and u_j^* denote the jth displacement and its constraint, respectively, and N is the total number of elements in the design domain. Note that the displacement constraint may be replaced by the mean compliance constraint (Chiandussi 2006; Huang and Xie 2009e). To a certain extent, the above optimization problem is equivalent to that in the previous chapters where the mean compliance is minimized subject to a given structural volume.

In order to solve this problem using the BESO method, the displacement constraint is added to the objective function by introducing a Lagrangian multiplier λ

$$
f_1(x) = \sum_{i=1}^{N} V_i x_i + \lambda(u_j - u_j^*)
$$

(6.2)

It is seen that the modified objective function is equivalent to the original one and the Lagrangian multiplier can be any constant if the displacement constraint is satisfied.

6.2.1 Sensitivity Analysis and Sensitivity Number

The derivative of the modified objective function $f_1(x)$ is

$$
\frac{df_1}{dx_i} = V_i + \lambda \frac{du_j}{dx_i}
$$

(6.3)

To calculate $\frac{du_j}{dx_i}$, we introduce a virtual unit load $\mathbf{f}_j$, in which only the corresponding jth component is equal to unity and all other components are equal to zero. Therefore,

$$
u_j = \mathbf{f}_j^T \mathbf{u}
$$

(6.4)

Based on the sensitivity analysis in previous chapters, we obtain

$$\frac{du_j}{dx_i} = -px_i^{p-1}\mathbf{u}_{ij}^T\mathbf{K}_i^0\mathbf{u}_i \tag{6.5}$$

where $\mathbf{u}_{ij}$ is found from the following adjoint equation

$$\mathbf{f}_j - \mathbf{K}\mathbf{u}_{ij} = \mathbf{0} \tag{6.6}$$

The above equation shows that $\mathbf{u}_{ij}$ is the virtual displacement vector of the ith element resulted from a unit virtual load $\mathbf{f}_j$. Note that the dimension of $\mathbf{u}_i$ and $\mathbf{u}_{ij}$ is the same as that of $\mathbf{u}$. However, all components in $\mathbf{u}_i$ and $\mathbf{u}_{ij}$ that are not related to element i are zero. Similarly, the dimensions of $\mathbf{K}_i^0$ are the same as those of $\mathbf{K}$ but all components in $\mathbf{K}_i^0$ that are not related to element i are zero. By substituting Equation (6.5) into (6.3), the derivative of the modified objective function becomes

$$\frac{df_1}{dx_i} = V_i - \lambda px_i^{p-1}\mathbf{u}_{ij}^T\mathbf{K}_i^0\mathbf{u}_i \tag{6.7}$$

When a uniform mesh is used (i.e. elements being of the same volume), the relative ranking of sensitivity of each element can be defined by the following sensitivity number

$$\alpha_i = -\frac{1}{\lambda p}\left(\frac{df_1}{dx_i} - V_i\right) = x_i^{p-1}\mathbf{u}_{ij}^T\mathbf{K}_i^0\mathbf{u}_i \tag{6.8}$$

In the BESO method, a structure is optimized by removing and adding elements. Only two discrete values are used, i.e. x_{min} for soft elements and 1 for solid elements. Therefore, the sensitivity numbers for solid and soft elements are explicitly expressed as

$$\alpha_i = \begin{cases} \mathbf{u}_{ij}^T\mathbf{K}_i^0\mathbf{u}_i & x_i = 1 \\ x_{min}^{p-1}\mathbf{u}_{ij}^T\mathbf{K}_i^0\mathbf{u}_i & x_i = x_{min} \end{cases} \tag{6.9}$$

6.2.2   Determination of Structural Volume

For the present optimization problem, the structural volume is to be minimized and determined according to the prescribed displacement constraint. With the sensitivity of the displacement u_j in Equation (6.5), the variation of the displacement due to changes in design variables can be estimated by

$$u_j^{k+1} \approx u_j^k + \sum_i \frac{du_j^k}{dx_i}\Delta x_i \tag{6.10}$$

where u_j^k and u_j^{k+1} denote the jth displacement in the current and next iterations respectively.

From the above equation, the threshold of the sensitivity number as well as the corresponding volume, V^c, can be easily determined by letting $u^{k+1} = u^*$. The procedure is similar to that described in Section 3.4. However, the resultant volume V^c may be much larger or far smaller than that of the current design. In order to have a gradual evolution of the topology, the following equation is adopted to determine the structural volume for the next iteration

$$V^{k+1} = \begin{cases} \max\left(V^k(1 - ER), V^c\right) & \text{when } V^k > V^c \\ \min\left(V^k(1 + ER), V^c\right) & \text{when } V^k \leq V^c \end{cases} \tag{6.11}$$

The above equation ensures that the volume change in each iteration be less than the prescribed evolutionary volume ratio, ER, which defines the maximum variation of the structural volume in a single iteration.

However the above procedure only applies to the soft-kill BESO method. In the hard-killed BESO method, the penalty exponent p is infinite and therefore the derivative of displacement in Equation (6.5) becomes infinite too for the solid element. An easy way to circumvent this problem is to use the following algorithm to determine the structural volume by comparing the current displacement with its constraint value (Huang and Xie 2009b)

$$V^{k+1} = \begin{cases} \max\left(V^k(1 - ER), \quad V^k\left(1 - \dfrac{u_j^k - u_j^*}{u_j^*}\right)\right) & \text{when } u_j^k > u_j^* \\ \min\left(V^k(1 + ER), \quad V^k\left(1 + \dfrac{u_j^* - u_j^k}{u_j^*}\right)\right) & \text{when } u_j^k \leq u_j^* \end{cases} \tag{6.12}$$

Numerical tests show that the above algorithm works well for the hard-kill BESO method. One such test is given below.

6.2.3   Examples

6.2.3.1   Example of Hard-kill BESO

The design domain of a short cantilever is shown in Figure 3.5. A point load of 100 N is applied downwards at the centre of the free end. The material has Young's modulus $E = 100$ GPa and Poisson's ratio $\nu = 0.3$. The mean compliance of the full design (C_0) is 1.157 Nmm. The design domain is divided into a uniform mesh of 50×80 four node plane stress elements. The mean compliance constraint $C^* = 2.083$ Nmm (or $u^* = 0.021$ mm at the loading point in the vertical direction) is 1.8 times that of the full design. The hard-kill BESO method with $ER = 1\%$ and $r_{min} = 3$ mm is used to search for an optimal design starting from the full design.

Figure 6.1 shows the evolution history of the topology. The final optimal topology is given in Figure 6.1(f). It is no surprise that the final topology has little difference from those seen previous chapters because the current optimization problem is somewhat equivalent to the previous optimization problem of minimizing the mean compliance subject to a volume constraint. Figure 6.2 shows evolution histories of the volume fraction and the mean compliance. If is noted that after 96 iterations the volume fraction converges to 44.3 % and the mean compliance converges to C^*.

6.2.3.2   Example of Soft-kill BESO

Consider the beam structure shown in Figure 3.9. A vertical load of 100 N is applied to the middle of the lower edge. Young's modulus $E = 100$ GPa and Poisson's ratio $\nu = 0.3$ are assumed. It is required that the maximum deflection of the beam should not exceed 0.2 mm

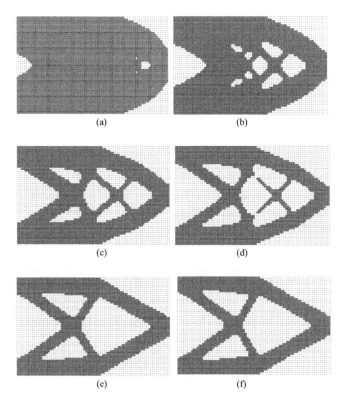

Figure 6.1 Evolution of topology: (a) iteration 10; (b) iteration 30; (c) iteration 50; (d) iteration 70; (e) iteration 80; (f) iteration 96.

under the given load. Due to the symmetry, the computation is performed for the right half of the design domain with 120×40 four node plane stress elements. The soft-kill BESO method with $ER = 2\%$, $x_{\min} = 0.001$, $p = 3$ and $r_{\min} = 1.5$ mm is used to search for an optimal design starting from the full design.

The final design shown in Figure 6.3 has 45 % volume of the initial full design. The maximum deflection is 0.1997 mm which is very close to the prescribed constraint limit, 0.2 mm. Evolution histories of the volume fraction and the constraint displacement are shown in Figure 6.4. Once again, both the volume fraction and the constraint displacement stably converge after 45 iterations.

6.3 Stiffness Optimization with an Additional Displacement Constraint

For the optimization problem discussed in the previous section, one must be careful in selecting the displacement constraint. Usually the chosen displacement relates directly to the overall

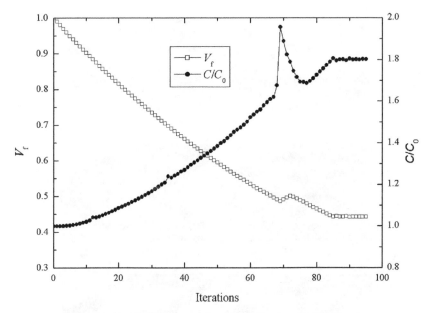

Figure 6.2   Evolution histories of mean compliance and volume fraction.

structural performance. For example, one may choose the maximum deflection at the loading point to be constrained. Otherwise, the previous optimization formulation could fail to find a reasonable design because the solution to an inappropriately defined problem may not exist.

In this section we shall consider a different optimization problem, i.e. maximizing stiffness with an additional displacement constraint. For example, a local displacement constraint may be imposed on the horizontal movement of the roller support shown in Figure 6.5. Such kind of constraint arises when the displacement at a specific location, not necessarily the loading point, is required to be within a prescribed limit.

The optimization problem with an additional displacement constraint has been studied for a truss system (Kočvara 1997). In topology optimization of geometrically nonlinear structures

Figure 6.3   Optimal design for displacement constraint $u^* = 2.0$ mm.

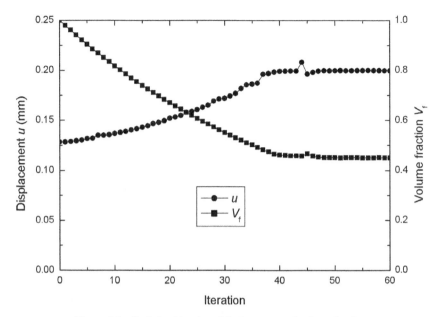

Figure 6.4 Evolution histories of displacement and volume fraction.

for compliant mechanisms, local displacements have been formulated into the objective functions rather than the constraints (Bendsøe and Sigmund 2003). For normal structures (as opposed to compliant mechanisms), a certain measure of the overall structural performance (such as stiffness) should be included in the objective function. A local displacement, on the other hand, is best to be considered as a constraint. The topology optimization problem of maximizing stiffness with a volume constraint and an additional displacement constraint can be

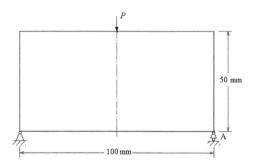

Figure 6.5 Design domain of a stiffness optimization problem with an additional displacement constraint $u_A \leq u_A^*$.

stated as

$$\text{Minimize } C = \frac{1}{2}\mathbf{f}^{\mathbf{T}}\mathbf{u}$$
$$\text{Subject to: } u_j \leq u_j^*$$
$$V^* - \sum_{i=1}^{N} V_i x_i = 0 \qquad (6.13)$$
$$x_i = x_{\min} \text{ or } 1$$

It is seen that the above optimization problem has multiple constraints. It is certainly possible to consider other optimization problems with a local displacement constraint in a similar manner, e.g. minimizing the structural volume against two displacement constraints in which one displacement is directly related to the overall structural performance. The BESO method developed below demonstrates the general principle for solving this type of topology optimization problems.

6.3.1 Sensitivity Number

We add the local displacement constraint to the objective function by introducing a Lagrangian multiplier λ. Thus, the modified objective function becomes

$$f_1(x) = \frac{1}{2}\mathbf{f}^{\mathbf{T}}\mathbf{u} + \lambda(u_j - u_j^*) = \frac{1}{2}\mathbf{u}^T\mathbf{K}\mathbf{u} + \lambda(u_j - u_j^*) \qquad (6.14)$$

The modified objective function is equivalent to the original one if the displacement is equal to its constraint value. Otherwise $\lambda = 0$ if $u_j < u_j^*$, which means the displacement constraint is already satisfied; and λ tends to infinity if $u_j > u_j^*$, which means we try to minimize the displacement u_j in order to satisfy the constraint in later iterations. Therefore, the Lagrangian multiplier is employed to act as a compromise between the objective function and the displacement constraint.

Using the adjoint method, the sensitivity of the modified objective function can be found as

$$\frac{df_1}{dx_i} = px_i^{p-1}\left(-\frac{1}{2}\mathbf{u}_i^T\mathbf{K}_i^0\mathbf{u}_i - \lambda\mathbf{u}_{ij}^T\mathbf{K}_i^0\mathbf{u}_i\right) \qquad (6.15)$$

Accordingly, the elemental sensitivity number, α_i, can be defined as

$$\alpha_i = -\frac{1}{p}\frac{df_1(x)}{dx_i} = x_i^{p-1}\left(\frac{1}{2}\mathbf{u}_i^T\mathbf{K}_i^0\mathbf{u}_i + \lambda\mathbf{u}_{ij}^T\mathbf{K}_i^0\mathbf{u}_i\right) \qquad (6.16)$$

As a result, the sensitivity numbers for solid and soft elements are expressed explicitly as

$$\alpha_i = \begin{cases} \dfrac{1}{2}\mathbf{u}_i^T\mathbf{K}_i^0\mathbf{u}_i + \lambda\mathbf{u}_{ij}^T\mathbf{K}_i^0\mathbf{u}_i & x_i = 1 \\[2mm] x_{\min}^{p-1}\left(\dfrac{1}{2}\mathbf{u}_i^T\mathbf{K}_i^0\mathbf{u}_i + \lambda\mathbf{u}_{ij}^T\mathbf{K}_i^0\mathbf{u}_i\right) & x_i = x_{\min} \end{cases} \qquad (6.17)$$

It is noted that the sensitivity numbers are dependent on the Lagrangian multiplier, λ, which needs to be determined first.

6.3.2 Determination of Lagrangian Multiplier

An appropriate value of λ can be determined when both the volume and displacement constraints are satisfied. For the ease of implementing the calculation, λ is defined as

$$\lambda = \frac{1 - w}{w} \tag{6.18}$$

where w is a constant ranging from w_{min}, e.g. 10^{-10}, to 1.

To find the appropriate value of w, the initial bound values of w are set to be $w_{lower} = w_{min}$ and $w_{upper} = 1$. The program starts from an initial guess of $w = 1$ and the sensitivity number is determined according to Equations (6.17) and (6.18). Thus, the threshold of sensitivity numbers can be determined when the volume constraint is satisfied by assuming that the density of element is x_{min} or 1 if the elemental sensitivity number is smaller or larger than the threshold accordingly.

Then the displacement in the next iteration, u_j^{k+1}, can be estimated using Equation (6.10). Thereafter, if $u_j^{k+1} > u_j^*$ we update w with a smaller value as follows

$$\hat{w} = \frac{w + w_{lower}}{2} \tag{6.19}$$

At the same time, we move the upper bound of w so that $w_{upper} = w$. On the other hand, if $u_j^{i+1} < u_j^*$, we update w with a larger value as follows

$$\hat{w} = \frac{w + w_{upper}}{2} \tag{6.20}$$

and the lower bound of w is updated so that $w_{lower} = w$.

With the updated $w = \hat{w}$, the above procedure is repeated until $w_{upper} - w_{lower}$ is less than 10^{-5}. Therefore, a total of 17 iterations are needed to obtain an accurate Lagrangian multiplier. This procedure is typically used for an additional constraint except for the volume constraint. The cost for computing the Lagrangian multiplier is negligible in the following examples because the calculation only needs to update the relative ranking of sensitivity numbers according to Equation (6.17) in each iteration. For problems with more than one additional constraint, multiple Lagrangian multipliers will need to be introduced and a more sophisticated algorithm should be developed to improve the computational efficiency.

6.3.3 Examples

6.3.3.1 Local Displacement Constraint at a Roller Support

In this example, stiffness optimization is carried out for the structure shown in Figure 6.5. In addition to a common volume constraint, a displacement constraint is imposed on the horizontal movement at the roller support. The vertical load $P = 100\,\text{N}$ is applied to the

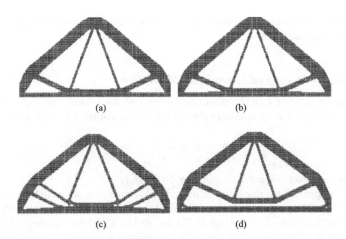

(a)   (b)

(c)   (d)

Figure 6.6   Various optimal designs: (a) without additional local displacement constraint; (b) $u_A^* = 1.4$ mm ; (c) $u_A^* = 1.2$ mm (d) $u_A^* = 1.0$ mm.

middle of the upper edge. Due to the symmetry, the computation is performed on the right half of the design domain with 100×100 four node plane stress elements. The material is assumed to have Young's modulus $E = 1$ GPa and Poisson's ratio $\nu = 0.3$. The volume constraint V^* is 30 % of the design domain. The soft-kill BESO method with $x_{min} = 0.001$, $p = 3$, $ER = 2\%$ and $r_{min} = 1.5$ mm is used.

The optimal topology without any displacement constraint is given in Figure 6.6(a) for the purpose of comparison. Its mean compliance is 191 Nmm and the horizontal movement of the roller support at point A is 1.43 mm. When the horizontal movement of the roller is constrained to be less than or equal to 1.4 mm, 1.2 mm or 1.0 mm, we obtain the topologies shown in Figures 6.6(b)–(d). Their mean compliances are 191 Nmm, 195 Nmm and 203 Nmm respectively. It is noted that there is an increase in the final mean compliance when a stricter displacement constraint is imposed.

Figure 6.7(a) presents evolution histories of the mean compliances for various displacement constraints. It shows that the objective function (mean compliance) is convergent at the final stage of the optimization process for all cases. Figure 6.7(b) gives evolution histories of the horizontal movement of the roller support. It is seen that the displacement converges to its constraint value at the final stage.

6.3.3.2   A Bridge-type Structure

In this example an optimization problem of designing a bridge is considered as shown in Figure 6.8. The design domain is a rectangle of the size $L \times H = 200$ mm $\times 40$ mm. The bottom deck of $L \times h = 200$ mm $\times 1$ mm is supported at two corners. The thickness of both the design domain and the deck is assumed to be 1 mm. A uniformly distributed load $p = 1$ N/mm is applied on the top surface of the deck. The design domain is divided

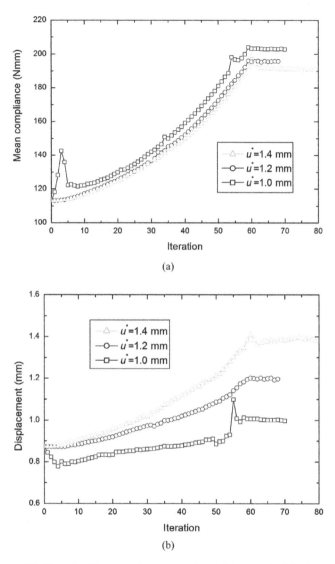

Figure 6.7   Evolution histories: (a) mean compliance; (b) constrained displacement.

into 200 × 40 four node plane stress elements and the nondesignable deck is divided into 200 beam elements. The nodes of the beam elements are connected to the corresponding nodes of the plate elements at the bottom side of the design domain. The material properties for the design domain are Young's modulus $E = 1$ GPa and Poisson's ratio $v = 0.3$, and the material properties for the deck are assumed to be Young's modulus $E = 100$ GPa and

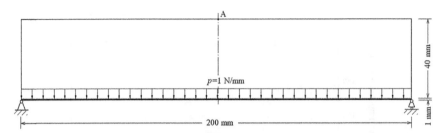

Figure 6.8   Design domain, loading and boundary conditions for a bridge-type structure.

Poisson's ratio $\nu = 0.3$. The volume constraint V^* is 50 % of the design domain. The soft-kill BESO method is used with $x_{min} = 0.001$, $p = 3$, $ER = 2\%$ and $r_{min} = 3.0$ mm.

Figure 6.9(a) presents the optimal design using the traditional BESO method without a local displacement constraint. It shows a tie-arch bridge with the deck suspended by a series of cables inclined at different angles. The mean compliance of this design is 93.5 Nmm and the vertical deflection at the middle of the upper edge (point A in Figure 6.8) is 2.7 mm. When the vertical deflection at A is constrained to be less than or equal to 2.4 mm and 2.0 mm, we obtain the topologies shown in Figures 6.9(b) and (c). With the additional displacement constraint, the deck is suspended by a series of cables that are almost parallel to each other.

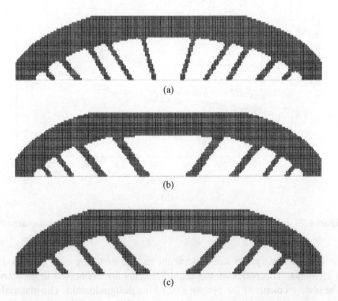

Figure 6.9   Various optimal designs: (a) without additional local displacement constraint; (b) $v_A^* = 2.4$ mm ; (c) $v_A^* = 2.0$ mm.

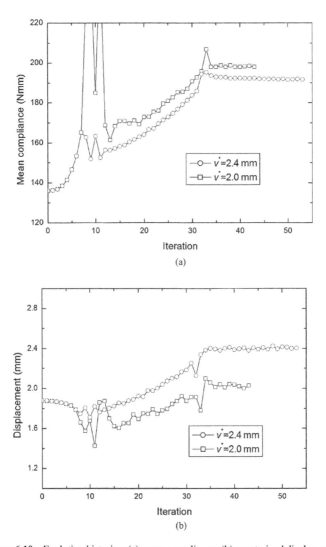

Figure 6.10   Evolution histories: (a) mean compliance; (b) constrained displacement.

The mean compliances are 95.8 Nmm and 99.0 Nmm for $v_A^* = 2.4$ mm and $v_A^* = 2.0$ mm respectively. Figure 6.10 shows evolution histories of the mean compliance and the constrained displacement. It is noted that both the mean compliance and the displacement are convergent at the final stage even though a few large jumps have occurred for $v_A^* = 2.0$ during early iterations.

6.4   Stiffness Optimization of Structures with Multiple Materials

In previous chapters we have discussed stiffness optimization of continuum structures with a single material. In recent years, composite materials are increasingly used in structural designs. In this section we extend the BESO method to stiffness optimization of structures with multiple materials. It is assumed that the elasticity moduli of different material phases are $E_1, E_2, \ldots, E_n$ (where $E_1 > E_2 \cdots > E_n$) respectively and the prescribed volume for each material j is V_j^*. The corresponding optimization problem can be stated as

$$\text{Minimize } C = \frac{1}{2}\mathbf{f}^T\mathbf{u}$$

$$\text{Subject to: } V_j^* - \sum_{i=1}^{N} V_i x_{ij} - \sum_{i=1}^{j-1} V_i^* = 0 \quad (j = 1, 2, \ldots, n - 1) \tag{6.21}$$

$$x_{ij} = x_{\min} \text{ or } 1$$

where the design variable x_{ij} for the ith element and the jth material is defined as follows

$$x_{ij} = \begin{cases} 1 & \text{for } E \geq E_j \\ x_{\min} & \text{for } E \leq E_{j+1} \end{cases} \tag{6.22}$$

6.4.1   Sensitivity Number

For structures composed of multiple materials, it is straightforward to interpolate the material properties between two neighbouring phases, e.g. E_j and E_{j+1}, as (Bendsøe and Sigmund 1999)

$$E(x_{ij}) = x_{ij}^p E_j + (1 - x_{ij}^p)E_{j+1} \tag{6.23}$$

where p is the penalty exponent.

Thus, the sensitivity number can be found through the sensitivity analysis with respect to the design variables x_{ij} as

$$\alpha_{ij} = -\frac{1}{p}\frac{\partial C}{\partial x_{ij}} = \frac{1}{2}x_{ij}^{p-1}\left(\mathbf{u}_i^T\mathbf{K}_i^j\mathbf{u}_i - \mathbf{u}_i^T\mathbf{K}_i^{j+1}\mathbf{u}_i\right) \tag{6.24}$$

where K_i^j and K_i^{j+1} denote the elemental stiffness matrices calculated from using E_j and E_{j+1} respectively. It should be noted that the sensitivity number α_{ij} is defined in the whole design domain even though it is only used for making adjustments between materials j and $j + 1$. The sensitivity number can be explicitly expressed as

$$\alpha_{ij} = \begin{cases} \dfrac{1}{2}\left[1 - \dfrac{E_{j+1}}{E_j}\right]\mathbf{u}_i^T\mathbf{K}_i^j\mathbf{u}_i & \text{for materials } 1, \cdots, j \\[3mm] \dfrac{1}{2}\dfrac{x_{\min}^{p-1}(E_j - E_{j+1})}{x_{\min}^p E_j + (1 - x_{\min}^p)E_{j+1}}\mathbf{u}_i^T\mathbf{K}_i^{j+1}\mathbf{u}_i & \text{for materials } j + 1, \cdots, n \end{cases} \tag{6.25}$$

where $\mathbf{K}_i^{j+1}$ is obtained from using $E(x_{\min}) = x_{\min}^p E_j + (1 - x_{\min}^p)E_{j+1}$ as the elasticity modulus.

For hard-kill BESO, p tends to infinity and the sensitivity number becomes

$$
\alpha_{ij} = \begin{cases} \dfrac{1}{2} \left[1 - \dfrac{E_{j+1}}{E_j} \right] \mathbf{u}_i^T \mathbf{K}_i \mathbf{u}_i & \text{for materials } 1, \cdots, j \\ 0 & \text{for materials } j+1, \cdots, n \end{cases} \tag{6.26}
$$

It is noted that there are $n-1$ groups of sensitivity numbers in the system to adjust the corresponding neighbouring materials. The BESO procedure for multiple material designs is similar to that for solid-void designs except that the sensitivity calculation and material adjustment must be carried out for each of the $n-1$ groups (Huang and Xie 2009a). BESO starts from a full design composed of material 1 and the evolution of topology begins with a prescribed evolutionary volume ratio ER, which is defined as the proportion of volume reduction of material 1 relative to the total volume of material 1 in the current design. At the same time, the volume of material 2 gradually increases until the volume constraint for material 2 is satisfied. Thereafter, the volume of material 2 remains constant and the volume of material 3 gradually increases until the volume constraint for material 3 is satisfied, and so on. Transition between materials 1 and 2 is carried out according to the target volume of material 1 for the next iteration and the sensitivity number α_{i1} (after the mesh-independency filter has been applied). Similarly, transition between materials 2 and 3 is carried out according to the target volume of material 2 for the next iteration and the filtered sensitivity number α_{i2}, and so on. The whole optimization procedure is stopped when the volume constraints for all materials are satisfied and the mean compliance becomes convergent.

6.4.2 Examples

6.4.2.1 Design with Two Nonzero Materials

In this example we perform topology optimization of the beam structure shown in Figure 2.10 using two nonzero materials (no void is allowed). The BESO method is used for the following four cases: (a) two materials with $E_1 = 1$ GPa and $E_2 = 0.1$ GPa, $p = 3$; (b) two materials with $E_1 = 1$ GPa and $E_2 = 0.1$ GPa, $p = $ infinity; (c) two materials with $E_1 = 1$ GPa and $E_2 = 0.2$ GPa, $p = 3$ and (d) two materials with $E_1 = 1$ GPa and $E_2 = 0.2$ GPa, $p = $ infinity. Note that when the penalty exponent p is infinite, the BESO method becomes a hard-kill approach. In all cases, the objective volume of material 1 is set to be 50 % of the design domain. The following BESO parameters are used: $ER = 2\%$, $r_{\min} = 3$ mm, and $x_{\min} = 1.0 \times 10^{-5}$ when $p = 3$.

Figure 6.11 gives the evolution histories of the mean compliance, the topology, and the volume fraction of material 1 for case (a). It shows that the present BESO method leads to a convergent solution for the objective function and the topology after 57 iterations. Figure 6.12 presents the final optimal topologies for all four cases. It is seen that the used value of the penalty exponent has little effect on the optimal topologies for multiple material even when p tends to infinity. The same behaviour has been observed for solid-void designs in previous chapters. The differences in the optimal topologies shown in Figure 6.12 are primarily due to the different materials used since the sensitivity numbers depend on the ratio of elasticity moduli of the two materials.

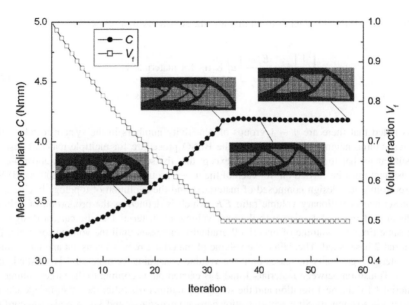

Figure 6.11 Evolution histories of mean compliance, volume fraction and topology of a bi-material structure when $E_1 = 1$ GPa, $E_2 = 0.1$ GPa and $p = 3.0$.

6.4.2.2 Design with Void and Two Nonzero Materials

We reconsider the above optimization problem by allowing void in the design as well as two materials with $E_1 = 1$ GPa and $E_2 = 0.1$ GPa. The Poisson's ratio is 0.3 for the two materials. The objective volumes of materials 1 and 2 are set to be 15 % and 25 % of the whole design domain. The above BESO method for multiple materials is used with $p = 3$ and $p = $ infinity. Soft elements with $x_{min} = 1.0 \times 10^{-5}$ is assumed when $p = 3$ while void elements are completely removed from the design when $p = $ infinity. Figure 6.13 shows the obtained optimal topologies for $p = 3$ and $p = $ infinity. There is no significant difference between the two optimal designs. It is also found that their mean compliances are almost identical (at 13.0 Nmm). The optimal topologies show a sandwich structure with a stiff skin (material 1) and a soft truss core (material 2). Figure 6.14 shows the evolution histories of the mean compliance, the topology, and the volume fractions of materials 1 and 2 when the hard-kill BESO method is used. Both the objective function and the topology start to converge to stable solutions after 130 iterations. It is worth noting, however, that the hard-kill BESO method ($p = $ infinity) is computationally more efficient (due to the complete removal of elements) than the soft-kill BESO method ($p = 3$).

6.5 Topology Optimization of Periodic Structures

Periodic structures such as the honeycomb core of a sandwich plate are widely used in structural designs due to their lightweight and ease of fabrication (Wadley *et al.* 2003). A lightweight cellular material usually possesses periodic microstructures. An inverse homogenization method has been proposed by Sigmund (1994) and Sigmund and Torquato (1997) using periodic

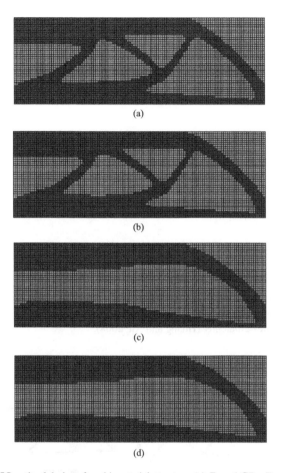

(a)

(b)

(c)

(d)

Figure 6.12   BESO optimal designs for a bi-material structure: (a) $E_1 = 1$ GPa, $E_2 = 0.1$ GPa, $p = 3$; (b) $E_1 = 1$ GPa, $E_2 = 0.1$ GPa, $p = $ infinity; (c) $E_1 = 1$ GPa, $E_2 = 0.2$ GPa, $p = 3$ and (d) $E_1 = 1$ GPa, $E_2 = 0.2$ GPa, $p = $ infinity.

boundary conditions to design cellular materials with prescribed effective properties. As a result, cellular materials with extreme macroscopic properties can found. However, the design of macro structures with periodic geometries, e.g. the core of a lightweight sandwich, for the mean compliance minimization is different from the pure material design of microstructures. Recently, Zhang and Sun (2006) have investigated scale-rated effects of the cellular material by combining the macroscopic design aimed at finding a preliminary layout of material and the refined design to determine locally the optimal material microstructure. The method can be directly applied to the design of periodic structures but the computational cost is high because it has to perform two finite element analyses in each iteration, one for the macroscale optimization problem and the other for the microscale sub-optimization problem. For the design of periodic structures, the macroscopic distribution of the designable material must be

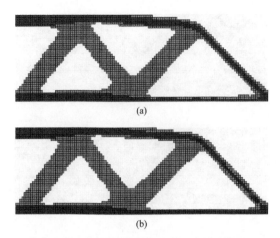

(a)

(b)

Figure 6.13   BESO optimal designs for a three-phase structure: (a) $p = 3.0$; (b) $p =$ infinity.

periodic even though the stress/strain distribution may not exhibit any periodic characteristics. Therefore, a general macroscopic optimization method with additional constraints on the periodicity of the geometry should be established in order to solve the optimization problems for periodic structures efficiently (Huang and Xie 2008).

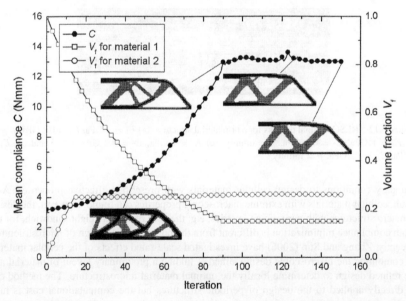

Figure 6.14   Evolution histories of mean compliance, volume fraction and topology of a three-phase structure when $p =$ infinity.

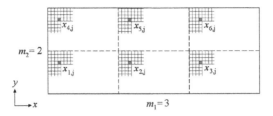

Figure 6.15 A two-dimensional design domain with 3×2 unit cells.

6.5.1 Problem Statement and Sensitivity Number

The objective of the present optimization problem is to find an optimal periodic topology of a structure with a given amount of material. Thus the resulting structure will have the maximum stiffness to carry the prescribed loading under given boundary conditions. To consider the periodicity of a structure, a two dimensional case is shown in Figure 6.15 as an example. The design domain is divided into $m = m_1 \times m_2$ unit cells, where m_1 and m_2 denote the number of unit cells along directions x and y respectively. It is noted that the special case of $m = 1 \times 1$ corresponds to the conventional topology optimization problem. We can formulate the optimization problem of minimizing the overall mean compliance using the binary design variables $x_{i,j}$, where i and j denote the cell number and the element number in the cell respectively (see Figure 6.15), as

$$\text{Minimize } C = \frac{1}{2}\mathbf{f}^T\mathbf{u}$$

$$\text{Subject to: } V^* - mV_i = 0$$

$$V_i = \sum_{j=1}^{n} V_{i,j}x_{i,j} \qquad (6.27)$$

$$x_{1,j} = x_{2,j} = \cdots = x_{m,j}$$

$$x_{i,j} = x_{\min} \text{ or } 1 (j = 1, 2, \cdots, n) (i = 1, 2, \cdots, m)$$

where n is the total number of elements in a unit cell, $V_{i,j}$ is the volume of the jth element in the ith unit cell and V^* is the prescribed total structural volume. The binary design variable $x_{i,j}$ declares the absence ($x_{\min}$) or presence (1) of an individual element. In the above equation, the condition $x_{1,j} = x_{2,j} = \cdots = x_{m,j}$ ensures that the status ($x_{\min}$ or 1) of elements at the corresponding locations in all unit cells should always be the same. This is how the periodicity of the design is maintained throughout the optimization process.

Due to the periodicity of the cells, the jth elements in all cells should be removed or added simultaneously. Therefore, the optimization process can be conducted in a representative unit cell, which can be selected from any unit cell by the user. The design variables of the representative unit cell are x_j. The sensitivity of jth element in the representative unit cell is defined by the variation of the overall mean compliance due to the combined change of the jth elements in all unit cells as

$$\frac{dC}{dx_j} = \sum_{i=1}^{m} \frac{dC}{dx_{i,j}} \frac{dx_{i,j}}{dx_j} = -\frac{1}{2} p \sum_{i=1}^{m} x_{i,j}^{p-1} \mathbf{u}_{i,j}^{T} \mathbf{K}_{i,j}^{0} \mathbf{u}_{i,j} \tag{6.28}$$

The elemental sensitivity number, which provides the relative ranking of the elemental sensitivity, can be written as

$$\alpha_j = -\frac{1}{p} \frac{dC}{dx_j} = \frac{1}{2} \sum_{i=1}^{m} x_{i,j}^{p-1} \mathbf{u}_{i,j}^{T} \mathbf{K}_{i,j}^{0} \mathbf{u}_{i,j} \tag{6.29}$$

When p tends to infinity, the sensitivity number for the hard-kill BESO method can be expressed as

$$\alpha_j = \begin{cases} \frac{1}{2} \mathbf{u}_{i,j}^{T} \mathbf{K}_{i,j}^{0} \mathbf{u}_{i,j} & x_j = x_{1,j} = \cdots = x_{m,j} = 1 \\ 0 & x_j = x_{1,j} = \cdots = x_{m,j} = 0 \end{cases} \tag{6.30}$$

The topology of a periodic structure can be defined by the representative unit cell, because the whole structure is divided into m identical cells. Therefore, BESO algorithms, such as that for filtering sensitivity numbers, can be applied to only the representative unit cell to save computational cost. However, it should be pointed out that the finite element analysis still needs to be performed on the whole structure because the stress/strain distributions in different cells are completely different.

6.5.2   Examples

6.5.2.1   A 2D Cantilever

Figure 6.16 shows a 2D rectangular domain with $L = 32$ and $H = 20$. The inner core with $H_1 = 16$ is the design domain. This optimization problem has been previously investigated

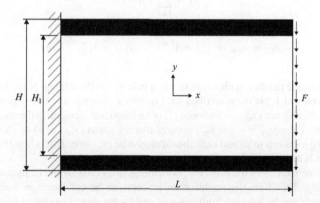

Figure 6.16 Design domain, loading and boundary conditions of the optimization problem in Zhang and Sun (2006).

by Zhang and Sun (2006). The plate is fixed on the left side and loaded vertically with $F = 100$ (force/length) on the right side. Young's modulus and Poisson's ratio of the material are $E = 1000$ and $v = 0.3$. The objective of the optimization problem is to find the optimal topology of the core with a volume constraint of 50 % of the total core space. In order to transfer the applied load properly, a nondesignable elastic thin strip is added to the right edge of the structure. Four different cases are studied using the hard-kill BESO method for $m = 2 \times 1$, 4×2, 8×4 and 16×8 respectively. To be consistent with the work by Zhang and Sun (2006), we arrange the unit cells in such a manner that they are symmetrical about the neutral axis in the second, third and fourth cases.

The filter radius $r_{\min}$ is selected to be 1.5, 1.0, 0.5 and 0.3 when $m = 2 \times 1$, 4×2, 8×4 and 16×8, respectively. BESO starts from the full design and gradually decreases the volume with the evolutionary volume ratio $ER = 2\%$ until the volume fraction constraint of 50 % is achieved. Then the volume is kept constant until the chosen convergence criterion, e.g. Equation (3.13), is satisfied. Figure 6.17 shows the evolution histories of the objective function (mean compliance), the volume fraction and the topology for $m = 2 \times 1$. It is seen that the solution is convergent towards the end of the optimization process. Figure 6.18 presents the final optimal topologies for all four cases. These designs are very similar to the optimal topologies obtained by Zhang and Sun (2006). The mean compliances of the BESO designs are 72 899.3, 73 843.3, 77 617.1 and 79 706.2 when $m = 2 \times 1$, 4×2, 8×4 and 16×8, respectively, which are much lower than the corresponding results of Zhang and Sun (2006), i.e. 82 530.6, 84 012.9, 88 308.3 and 90 547.5, respectively. These differences may be attributed, to a large extent, to the over-estimation of the strain energy of the soft elements when the penalty exponent p is greater than 1.

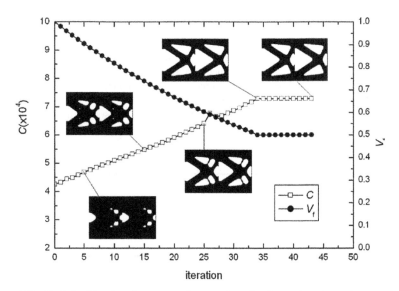

Figure 6.17 Evolution histories of mean compliance, volume fraction and topology when $m = 2 \times 1$.

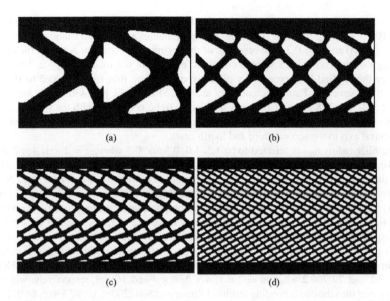

Figure 6.18 Optimal designs for various periodic constraints: (a) $m = 2 \times 1$; (b) $m = 4 \times 2$; (c) $m = 8 \times 4$ and (d) $m = 16 \times 8$.

6.5.2.2 A 2D Sandwich Structure

The present BESO method can be applied to the design of the cellular core in a sandwich structure. As illustrated in Figure 6.19, the sandwich structure is fixed at both ends of the skins. The designable core is a rectangle of 160 mm in length and 40 mm in width, which is divided into 320×80 four node plane stress elements. The two skins, which are 1 mm in thickness, are divided into 320 beam elements for each skin. It is assumed that the skins and the core are tied together. A vertical point force $P = 1$ N is applied at the middle point of the top skin. The materials of the skins and the core are assumed to have Young's modulus 100 GPa and Poisson's ratio 0.3, and Young's modulus 1GPa and Poisson's ratio 0.3 respectively.

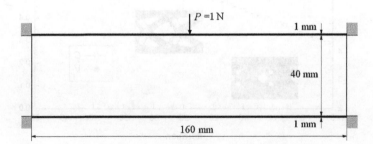

Figure 6.19 Design domain, loading and boundary conditions of a 2D sandwich structure.

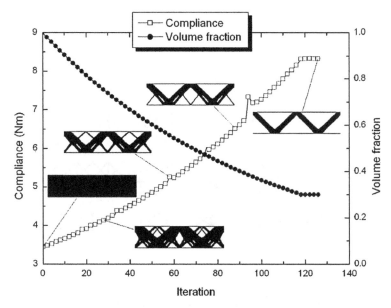

Figure 6.20 Evolution histories of mean compliance, volume fraction and topology of the sandwich structure with $m = 2 \times 1$ when BESO starts from the initial full design.

The volume constraint is 30 % of the core space. The filter radius and evolutionary volume ratio are selected to be $r_{min} = 2$ mm and $ER = 1$ %. The hard-kill BESO method is used.

Figure 6.20 shows the evolution histories of the objective function (mean compliance), the volume fraction and the topology for $m = 2 \times 1$ when the BESO starts from the initial full design. It is seen that all topologies satisfy the prescribed periodic constraint. The mean compliance of the final topology is 8.33 Nm. Table 6.1 presents the optimal topologies and their mean compliances for various numbers of unit cells. A typical unit cell is given inside dashed lines except for $m = 1 \times 1$. Figure 6.21 plots the values of the final mean compliance against the number of unit cells. It is seen that the mean compliance increases with the number of unit cells. When the number of unit cells increases, the total number of independent design variables decreases. Therefore, the solution of the conventional BESO method corresponding to the special case of $m = 1 \times 1$ has the lowest mean compliance among all optimal designs. On the other hand, the optimal topology depends on the aspect ratio of the unit cell. For example, the optimal topologies for $m = 2 \times 1$ and $m = 1 \times 2$ are totally different, even though their total numbers of unit cells are equal.

The BESO method may also start from an initial guess design. Such an initial design does not have to be an 'intelligent' guess or a near-optimum. Indeed, it could be a highly nonoptimal design. As an example, Figure 6.22 presents the evolution histories of the mean compliance, the volume fraction and the topology for $m = 2 \times 1$ when the hard-kill BESO starts from the initial guess design shown at the top left corner. The initial guess design bears no resemblance to the final optimal topology. At the beginning of the evolution process, both the mean compliance and the volume decrease, which implies that the overall stiffness of the

Table 6.1 Optimal designs and their mean compliances for the sandwich structure under various periodic constraints.

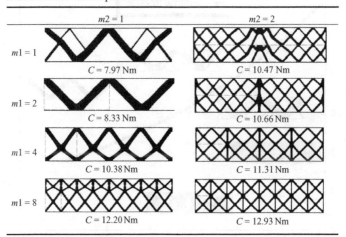

	$m2 = 1$	$m2 = 2$
$m1 = 1$	$C = 7.97$ Nm	$C = 10.47$ Nm
$m1 = 2$	$C = 8.33$ Nm	$C = 10.66$ Nm
$m1 = 4$	$C = 10.38$ Nm	$C = 11.31$ Nm
$m1 = 8$	$C = 12.20$ Nm	$C = 12.93$ Nm

structure is increased due to the optimization algorithm despite the fact that the volume of the structure is reduced. The volume fraction remains constant after the 7th iteration and the compliance continues to decrease until it converges to a constant value, 8.33 Nm, which is the same as that of the solution shown in Figure 6.20. One advantage of starting from an initial guess design with a volume close or equal to the target volume is that, while the volume is kept constant, the improvement in the objective function can be clearly seen in its evolution history. Another advantage is that this procedure is computationally more efficient as only a portion of all elements in the full design domain needs to be included in the finite element analyses. However, it should be pointed out that BESO starting from an initial guess design may sometimes converge to a local optimum (Huang and Xie 2007).

6.5.2.3 A Bridge-type Structure

Figure 6.23 shows an optimization problem of designing a bridge-type structure. The design domain is a rectangle of length $L = 240$ and depth $H = 60$ (thickness $t = 1$). The deck at the bottom with length $L = 240$, depth $h = 5$ (and thickness $t = 1$) is supported at two corners. A vertical load $P = 100$ is applied to the mid point of the deck. The design domain is divided into 240×60 four node plane stress elements and the nondesignable deck is modelled with 240 beam elements. The nodes of the beam elements are connected to those of the plate elements at the bottom side of the design domain. It is assumed that the materials for the design domain and the deck are the same with Young's modulus $E = 2000$ and Poisson's ratio $\nu = 0.3$. The volume fraction constraint is 30 % of the design domain. The hark-kill BESO method with $ER = 2$ % is used in this example.

Figure 6.24(a) shows the optimal design for $m = 1 \times 1$, which is the same as the conventional optimal design without any periodic constraint. The mean compliance of this design is 1.12. Figures 6.24(b) and (c) show optimal designs for $m = 4 \times 1$ and $m = 6 \times 1$ respectively.

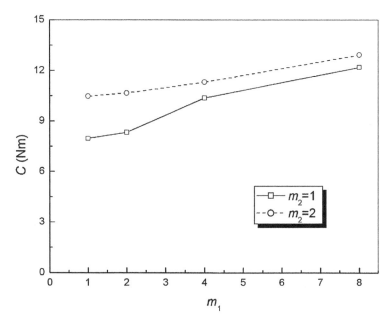

Figure 6.21   Effect of the number of unit cells on the mean compliance.

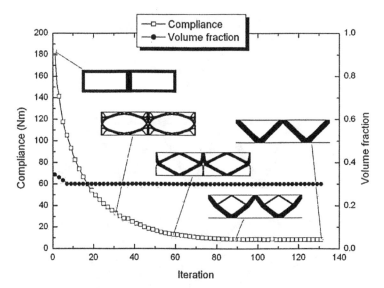

Figure 6.22   Evolution histories of mean compliance, volume fraction and topology of the sandwich structure with $m = 2 \times 1$ when BESO starts from an initial guess design.

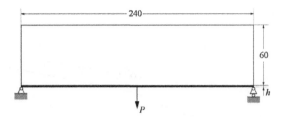

Figure 6.23 Design domain, loading and boundary conditions of a 2D bridge-type structure.

Their mean compliances are 1.53 and 1.78, which are higher than that of the conventional design. Similar to the example discussed in the previous section, the mean compliance increases with the total number of unit cells.

To study the influence of the deck stiffness, the above problem is reconsidered by increasing the deck depth h from 5 to 50. The optimal design for $m = 4 \times 1$ is shown in Figure 6.25. The result is significantly different from the design shown in Figure 6.24(b). This demonstrates that the optimal topology may depend on the stiffness of nondesignable parts.

6.5.2.4 A 3D Sandwich Structure

The present BESO method can be used to design 3D periodic structures. Figure 6.26(a) shows a sandwich cantilever subjected to four vertical loads ($P = 1$). The designable core of the

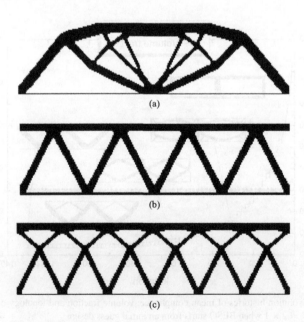

Figure 6.24 Various optimal designs for the bridge-type structure: (a) $m = 1 \times 1$, $C = 1.12$; (b) $m = 4 \times 1$, $C = 1.53$; (c) $m = 6 \times 1$, $C = 1.78$.

Figure 6.25 Optimal design for a bridge-type structure with a stronger deck ($h = 50$) when $m = 4 \times 1$.

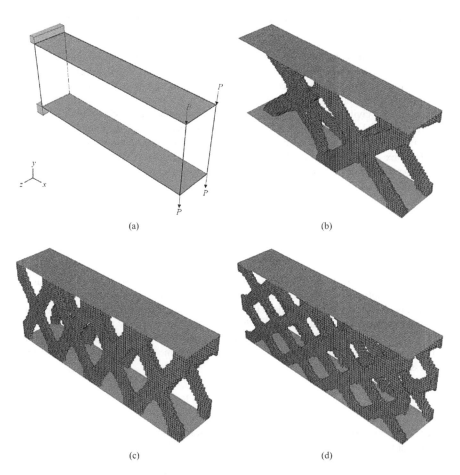

(a)

(b)

(c)

(d)

Figure 6.26 Topology optimization of 3D sandwich structure: (a) design domain, loading and boundary conditions; (b) $m = 1 \times 1 \times 1$, $C = 40.9$; (c) $m = 4 \times 1 \times 1$, $C = 45.4$; (d) $m = 4 \times 2 \times 1$, $C = 49.5$.

size $100 \times 40 \times 20$ is divided into $100 \times 40 \times 20$ eight-node cubic elements. Each of the nondesignable skins at the top and the bottom is of unit thickness and is divided into 100×20 four node plate elements. Young's modulus $E = 1$ and Poisson's ratio $\nu = 0.3$ are assumed for the core, and $E = 100$ and $\nu = 0.3$ for the skins. The objective is to find the optimal periodic layout of the core by minimizing the mean compliance of the structure subject to a prescribed volume constraint which is set to be 10 % of the designable domain. The hard-kill BESO method with $ER = 2\%$ is used in this example.

Figures 6.26(b), (c) and (d) show the optimal topologies for $m = 1 \times 1 \times 1$, $m = 4 \times 1 \times 1$ and $m = 4 \times 2 \times 1$. Their mean compliances are 40.9, 45.4 and 49.5 respectively. Again it is noted that the mean compliance increases with the total number of unit cells.

6.6 Topology Optimization of Structures with Design-dependent Gravity Load

Compared with the compliance minimization of structures with fixed loads, topology optimization of structures subjected to design-dependent loads, such as self-weight or any kind of body forces from centrifugal acceleration in rotating machines, has not been studied extensively. Gravity load is of significant importance in many engineering problems such as the design of large civil structures. Topology optimization of structures with gravity load faces several numerical difficulties, e.g. the nonmonotonous behaviour of the compliance and the parasitic effect of low density elements when the power law material model is used (Bruyneel and Duysinx 2005). Using a modified ESO procedure, Yang *et al.* (2005) and Ansola *et al.* (2006) have studied the topology optimization of structures under self-weight. An obvious shortcoming of the ESO procedure is that the material cannot be recovered once it is wrongly removed. In the following sections, the problem of topology optimization of structures with gravity load is studied using an extended BESO method (Huang and Xie 2009c).

6.6.1 Sensitivity Analysis and Sensitivity Number

The basic formulation of the topology optimization problem may also consist of minimizing the mean compliance of the body force subject to a constraint on the structural volume as given in Equation (4.1). Different from the topology optimization problem for fixed external forces, here the applied force, $\mathbf{f}$, includes the design-dependent gravity load.

The adjoint method can be used to determine the sensitivity of the displacement and the force by introducing a Lagrangian multiplier vector λ. The modified objective function can be expressed as

$$C = \frac{1}{2}\mathbf{f}^T\mathbf{u} + \lambda^T(\mathbf{f} - \mathbf{K}\mathbf{u}) \tag{6.31}$$

where the term $\lambda^T(\mathbf{f} - \mathbf{K}\mathbf{u})$ is equal to zero and therefore the modified objective function is equivalent to the original one. The variation of the modified objective function is

$$\frac{dC}{dx_i} = \frac{1}{2}\frac{d\mathbf{f}^T}{dx_i}\mathbf{u} + \frac{1}{2}\mathbf{f}^T\frac{d\mathbf{u}}{dx_i} + \lambda^T\left(\frac{d\mathbf{f}}{dx_i} - \frac{d\mathbf{K}}{dx_i}\mathbf{u} - \mathbf{K}\frac{d\mathbf{u}}{dx_i}\right) + \frac{d\lambda^T}{dx_i}(\mathbf{f} - \mathbf{K}\mathbf{u}) \tag{6.32}$$

where the first term denotes the variation of the force vector with respect to the design variable, x_i, and the last term is equal to zero because of the equilibrium equation $\mathbf{Ku} = \mathbf{f}$. The above equation can be rewritten as

$$\frac{dC}{dx_i} = \frac{1}{2}\frac{d\mathbf{f}^{\mathrm{T}}}{dx_i}(\mathbf{u} + 2\lambda) + \left(\frac{1}{2}\mathbf{f}^{\mathrm{T}} - \lambda^{\mathrm{T}}\mathbf{K}\right)\frac{d\mathbf{u}}{dx_i} - \lambda^{\mathrm{T}}\frac{d\mathbf{K}}{dx_i}\mathbf{u} \tag{6.33}$$

To eliminate the unknown $\frac{d\mathbf{u}}{dx_i}$ from the sensitivity expression, let λ be

$$\lambda = \frac{1}{2}\mathbf{u} \tag{6.34}$$

Thus, the sensitivity of the objective function is

$$\frac{dC}{dx_i} = \frac{d\mathbf{f}^{\mathrm{T}}}{dx_i}\mathbf{u} - \frac{1}{2}\mathbf{u}^{\mathrm{T}}\frac{d\mathbf{K}}{dx_i}\mathbf{u} \tag{6.35}$$

It is seen that the sensitivity of the mean compliance can be both positive and negative. It may even change sign when the value of the design variable is changed. Therefore, the mean compliance exhibits nonmonotonous behaviour.

When the power law material interpolation scheme Equation (4.2) is used, the ratio between the first and second terms of Equation (6.35) becomes unbounded in the low-density regions. Due to this reason, it has been demonstrated that it is almost impossible to obtain a 0/1 design using the power law material interpolation scheme (Bruyneel and Duysinx 2005). Therefore, we consider an alternative interpolation scheme proposed by Stolpe and Svanberg (Stolpe and Svanberg 2001) which overcomes the above shortcoming of the power law material model. The density and Young's modulus of the material model for a 0/1 design are given as

$$\rho_i = x_i \rho^0$$
$$E_i = \frac{x_i}{1 + q(1 - x_i)}E^0 \tag{6.36}$$

where ρ^0 and E^0 denote the density and Young's modulus of the solid material, and q is the penalty factor which is larger than 0 for topology optimization problems.

A structure is usually modelled using finite elements. As an example, let us consider a model with four node quadrilateral elements. The elemental load vector due to its gravity is as follows (assuming that the gravity is aligned with the global y direction)

$$\mathbf{f}_i = V_i \rho_i g \bar{\mathbf{f}} = V_i \rho_i g \left\{0, -\frac{1}{4}, 0, -\frac{1}{4}, 0, -\frac{1}{4}, 0, -\frac{1}{4}\right\}^T \tag{6.37}$$

where g is the gravity acceleration. A similar gravity load can be found for an eight node brick element for 3D structures by assigning 1/8 of the self-weight of the element to each node. From Equations (6.35), (6.36) and (6.37), we obtain the sensitivity of the mean compliance as

$$\frac{dC}{dx_i} = V_i \rho^0 g \bar{\mathbf{f}}^T \mathbf{u}_i - \frac{1 + q}{2\left[1 + q(1 - x_i)\right]^2}\mathbf{u}_i^T \mathbf{K}_i^0 \mathbf{u}_i \tag{6.38}$$

It is noted that the ratio between the first and second terms of the above equation becomes a finite constant when $x_i = x_{min}$. The sensitivity number can be defined as

$$
\alpha_i = -\frac{1}{q+1}\frac{dC}{dx_i} =
\begin{cases}
-\dfrac{V_i \rho^0 g}{q+1}\bar{\mathbf{f}}^T \mathbf{u}_i + \dfrac{1}{2}\mathbf{u}_i^T \mathbf{K}_i^0 \mathbf{u}_i & x_i = 1 \\[4mm]
-\dfrac{V_i \rho^0 g}{q+1}\bar{\mathbf{f}}^T \mathbf{u}_i + \dfrac{1}{2\left[1+q(1-x_{min})\right]^2}\mathbf{u}_i^T \mathbf{K}_i^0 \mathbf{u}_i & x_i = x_{min}
\end{cases}
\tag{6.39}
$$

To minimize the mean compliance, we update design variables by switching x_i from 1 to x_{min} for elements with the lowest sensitivity numbers and from x_{min} to 1 for elements with the highest sensitivity numbers. Similar to the power law material interpolation scheme, a large penalty factor q should be selected so that the BESO approach using discrete design variables (1 and x_{min}) would converge to a stable 0/1 design.

From Equation (6.36) it is noted that Young's modulus of soft elements becomes zero when the penalty factor q tends to infinity. In such a case, the soft element can be considered to be totally removed from the structure because the elemental stiffness is zero. Therefore, when q is infinite the sensitivity number reduces to the following simple expression

$$
\alpha_i =
\begin{cases}
\dfrac{1}{2}\mathbf{u}_i^T \mathbf{K}_i^0 \mathbf{u}_i & x_i = 1 \\[4mm]
0 & x_i = 0
\end{cases}
\tag{6.40}
$$

where $x_i = x_{min}$ is replaced by $x_i = 0$ because the soft element with extremely small stiffness is equivalent to a void element. The above equation implies that the hard-kill BESO method for structures with design-dependent gravity load is exactly the same as that for structures with fixed external forces.

It should be pointed out that the nonmonotonous behaviour of the objective function depends on the value of the penalty factor q. With a larger penalty factor, the nonmonotonous behaviour of the objective function becomes weaker. In the extreme case of an infinite penalty factor, the nonmonotonous behaviour disappears completely as is evident from Equation (6.40). Different from the topology optimization of structures with fixed external forces, here the selection of the penalty factor does affect the relative ranking of the sensitivity numbers of solid elements. Therefore, the total elimination of the first term of Equation (6.39) in the hard-kill BESO method may result in a different solution from that of the soft-kill BESO method. Nevertheless, it is still worth exploring possibilities of the hard-kill BESO method due to its high computational efficiency.

6.6.2 Examples

6.6.2.1 An Arch under Gravity Load

A rectangular design domain of 1 m $\times$ 0.5 m is simply supported at the bottom corners as illustrated in Figure 6.27. The structure is under gravity load only. The amount of available material is set to be 15 % of the design domain. The material is assumed to have the following properties: Young's modulus 200 GPa, Poisson's ratio 0.3 and the density 78 kg/m^3. Due to symmetry, only half of the design domain is analysed using a mesh of 100 $\times$ 50 four node

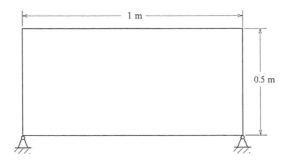

Figure 6.27 Design domain and boundary conditions of a structure under gravity load.

plane stress elements. Upon completion of the evolution process, the half model is mirrored
to obtain the whole structure. In this example, both soft-kill and hard-kill BESO methods
are used. The parameters for the soft-kill BESO are $ER = 2\%$, $AR_{max} = 2\%$, $q = 5$ and
$r_{min} = 30$ mm. The parameters for the hard-kill BESO are the same as those for the soft-kill
BESO except for the inherent infinite penalty factor q.

The evolutionary optimization is carried out based on the sensitivity numbers derived above.
Initially, all elements in the design domain are assigned to be solid. The total volume of the
structure gradually decreases and then is kept constant when the prescribed volume constraint
is achieved. Figure 6.28 shows the evolution histories of the mean compliance for the soft-kill

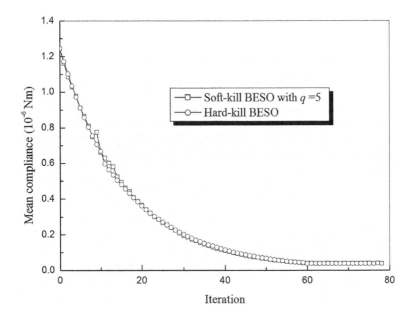

Figure 6.28 Evolution histories of mean compliance for soft-kill and hard-kill BESO methods.

(a)   (b)

Figure 6.29 Optimal designs from two different BESO methods: (a) soft-kill BESO with $q = 5$, $C = 3.82 \times 10^{-8}$ Nm; (b) hard-kill BESO method, $C = 3.81 \times 10^{-8}$ Nm.

and hard-kill BESO methods. It is seen that the mean compliance gradually decreases and then converges to a constant value in less than 80 iterations. The evolution histories of the mean compliance for the two BESO methods are almost identical. The hard-kill BESO method even gives a more stable evolution history of the mean compliance and reaches a convergent solution slightly earlier although its computation time is much less than that of the soft-kill BESO method.

Figure 6.29 shows the optimal topologies obtained from the soft-kill and hard-kill BESO methods. In both cases, the optimization process yields an arch structure which spans the two ends to support its self-weight effectively and efficiently. The resulted optimal shapes from the soft-kill and hard-kill BESO methods have little difference. The objective functions are 3.82×10^{-8} Nm and 3.81×10^{-8} Nm for the soft-kill and hard-kill designs respectively, which are also very close.

6.6.2.2 A Beam under Both Gravity and Concentrated Loads

In this example, the classic MBB beam shown in Figure 6.30 is considered. In additional to a concentrated load applied to the top of the beam, the self-weight of the structure is included. The dimensions of the design domain are 20 m × 5 m × 1 m. Due to symmetry, only half of the design domain is analysed using 100 × 50 four node plane stress elements. The volume constraint is set to be 40 % of the whole design domain. Young's modulus $E = 200$ GPa, Poisson's ratio $v = 0.3$ and density $\rho = 78$ kg/m^3 are assumed. Therefore, the total weight

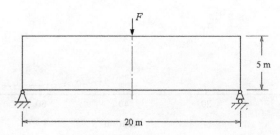

Figure 6.30 Design domain, loading and boundary conditions of a beam.

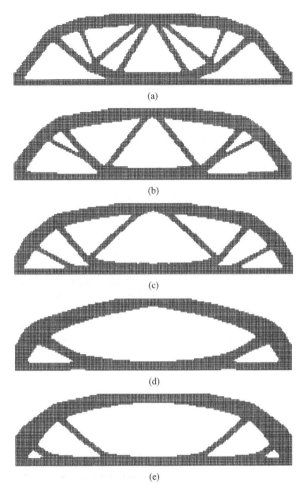

Figure 6.31 Optimal designs for different ratios between the concentrated load and the self-weight: (a) no self-weight: (b) $F = 100\%$ of self-weight, $C = 0.282$ Nm; (c) $F = 50\%$ of self-weight, $C = 0.131$ Nm; (d) $F = 10\%$ of self-weight, $C = 0.048$ Nm; (e) self-weight only, $C = 0.034$ Nm.

of the final design is 3120 kg. The soft-kill BESO method with $ER = 2\%$, $AR_{max} = 2\%$, $q = 5$, $x_{min} = 0.001$ and $r_{min} = 0.3$ m is used in this example.

When the self-weight is not taken into account, the magnitude of the concentrated load F has no effect on the resulting optimal topology shown in Figure 6.31(a). In this case, 76 iterations are needed to reach the convergent solution. It shows that the soft-kill BESO method utilizing the alternative material interpolation model given in Equation (6.36) is also efficient to solve the optimization problem of minimizing the mean compliance of structures with fixed external loads.

Figure 6.32   Optimal design of a structure with self-weight only using the hard-kill BESO method.

When the self-weight is taken into account, several magnitudes of the concentrated load
F are tested. The optimal designs and their corresponding mean compliances are shown in
Figure 6.31. The total numbers of iterations are 57, 58, 59 and 57 respectively before the
converged solutions shown in Figures 6.31(b)–(e) are obtained. It is seen that the resulting
topology depends on the ratio between the concentrated load and the structural self-weight.
Such behaviour has also been observed by Bruyneel and Duysinx (2005). When the applied
concentrated load decreases, the shape of the structure tends to become a tie-arch, which
makes sense from an engineering point of view.

To further test the effectiveness of the hard-kill BESO method for the nonmonotonous
problem, we apply it to the above optimization problem with self-weight only. The same
BESO parameters are used, except for the penalty factor which is now infinite. Figure 6.32
shows the optimal topology from the hard-kill BESO method, which is similar to that of the
soft-kill BESO method. The evolution histories of the mean compliances for both methods
are given in Figure 6.33. It is seen that the soft-kill BESO method produces a slightly better
solution than that of the hard-kill BESO method. The mean compliance of the hard-kill design

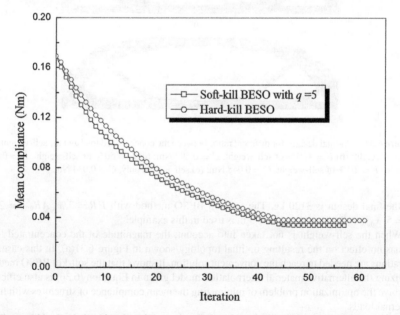

Figure 6.33   Evolution histories of mean compliances for soft-kill and hard-kill BESO methods.

is 0.037 Nm, which is about 9 % higher than that of the soft-kill design. It is interesting to note that a plate of uniform thickness of 0.4 m which has the same weight as that of the optimal designs has a mean compliance is 0.068 Nm. Compared with the uniform plate design, the optimal topologies from both BESO methods provide significant improvements in the overall structural stiffness.

6.6.2.3 A 3D Structure under Gravity Load

The above examples demonstrate that the hard-kill BESO method is able to obtain optimal solutions for structures with design-dependent gravity load, even though the mean compliance may sometimes be slightly higher than that of the soft-kill solution. However, the main advantage of the hard-kill BESO is its computational efficiency, especially for complex 3D structures. Furthermore, the original nonmonotonous optimization problem becomes monotonous when an infinity penalty factor is applied. Therefore, it is worthwhile testing the hard-kill BESO method on a 3D structure under the gravity load.

The structure shown in Figure 6.34(a) is simply supported at the four lower corners and is loaded with its self-weight only. Young's modulus $E = 200$ GPa, Poisson's ratio $v = 0.3$ and density $\rho = 78$ kg/m^3 are assumed. Due to symmetry in two directions, only a quarter of the design domain is analysed using $40 \times 50 \times 40$ eight node brick elements. The volume constraint is set to be 5 % of the whole design domain. The hard-kill BESO method with $ER = 4\%$, $AR_{max} = 4\%$ and $r_{min} = 0.3$ m is used in this example.

The obtained optimal topology is shown in Figure 6.34(b). The final structure consists of four interconnected inclined arches. Figure 6.35 shows the evolution histories of the mean compliance and the volume fraction. The mean compliance stably converges to 782.5 Nm after 82 iterations while the volume fraction reaches its constraint value of 5 %.

6.7 Topology Optimization for Natural Frequency

Frequency optimization is of great importance in many engineering fields, e.g. aerospace and automotive industries. Compared with the extensive research on stiffness optimization, there has been much less work concerned with topology optimization for natural frequency. Tenek

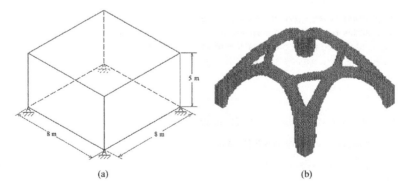

(a)   (b)

Figure 6.34 Topology optimization of a 3D structure under gravity load: (a) design domain and boundary conditions; (b) optimal design.

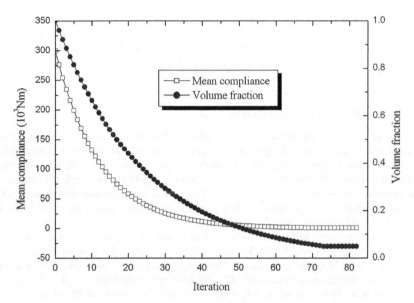

Figure 6.35   Evolution histories of mean compliance and volume fraction.

and Hagiwara (1994) and Ma *et al.* (1995) have used the homogenization method and Kosaka and Swan (1999), Pedersen (2000) and Du and Olhoff (2007) have used the SIMP method for frequency optimization. However, it has been demonstrated that the SIMP model is unsuitable for frequency optimization due to artificial localized modes in low density regions (Pedersen 2000). Therefore, a modified SIMP model using a discontinuous function has been proposed and used to solve frequency optimization problems (Pedersen 2000; Du and Olhoff 2007). Because the BESO method uses two discrete design variables x_{min} and 1 only, it is impossible to apply the above discontinuous material interpolation scheme.

There has been some research on ESO/BESO methods for frequency optimization (e.g. Xie and Steven 1996; Yang *et al.* 1999). As a result of having elements eliminated totally from the design domain, the original hard-kill ESO/BESO methods effectively prevent localized vibration modes from occurring. However, various deficiencies of the original ESO/BESO methods as identified in the previous chapters also exist in frequency optimization problems. In the following section a new soft-kill BESO method for frequency optimization is developed (Huang *et al.* 2009).

6.7.1   Frequency Optimization Problem and Material Interpolation Scheme

6.7.1.1   Frequency Optimization Problem

In finite element analysis, the dynamic behaviour of a structure can be represented by the following general eigenvalue problem

$$(\mathbf{K} - \omega_j^2 \mathbf{M})\mathbf{u}_j = \mathbf{0} \tag{6.41}$$

where $\mathbf{K}$ is the global stiffness matrix, $\mathbf{M}$ is the global mass matrix, ω_j is the jth natural frequency and $\mathbf{u}_j$ is the eigenvector corresponding to ω_j. ω_j and $\mathbf{u}_j$ are related to each other by the following Rayleigh quotient

$$\omega_j^2 = \frac{\mathbf{u}_j^T \mathbf{K} \mathbf{u}_j}{\mathbf{u}_j^T \mathbf{M} \mathbf{u}_j} \tag{6.42}$$

Here, we consider the topology optimization problem of maximizing the natural frequency ω_j. For a solid-void design, the optimization problem can be stated as

$$\text{Maximize } \omega_j$$
$$\text{Subject to: } V^* - \sum_{i=1}^{N} V_i x_i = 0 \tag{6.43}$$
$$x_i = x_{\min} \text{ or } 1$$

where V_i is the volume of the ith element and V^* is the prescribed total structural volume.

6.7.1.2 An Alternative Material Interpolation Scheme

To obtain the gradient information of the design variable, it is necessary to interpolate the material properties between $x_{\min}$ and 1. A popular material interpolation scheme is to apply the power law penalization model given in Equation (4.2) to the stiffness and a linear interpolation model to the density (i.e., $\rho_i = x_i \rho^0$). However, such a scheme results in numerical difficulties (Pedersen 2000). The main problem is that the extremely high ratio between mass and stiffness for small values of x_i (when the penalty exponent is greater than 1) causes artificial localized vibration modes in the low density regions. One idea to avoid this problem is to keep the ratio between mass and stiffness constant when $x_i = x_{\min}$ by requiring that

$$\rho(x_{\min}) = x_{\min} \rho^0$$
$$E(x_{\min}) = x_{\min} E^0 \tag{6.44}$$

where ρ^0 and E^0 denote the density and Young's modulus of the solid material. Therefore, an alternative material interpolation scheme can be expressed as

$$\rho(x_i) = x_i \rho^0$$
$$E(x_i) = \left[\frac{x_{\min} - x_{\min}^p}{1 - x_{\min}^p}(1 - x_i^p) + x_i^p \right] E^0 \quad (0 < x_{\min} \le x_i \le 1) \tag{6.45}$$

Figure 6.36 illustrates the above model for several values of p when $x_{\min} = 0.01$.

From Equation (6.45), the derivatives of the global mass matrix $\mathbf{M}$ and stiffness matrix $\mathbf{K}$ can be obtained

$$\frac{\partial \mathbf{M}}{\partial x_i} = \mathbf{M}_i^0$$
$$\frac{\partial \mathbf{K}}{\partial x_i} = \frac{1 - x_{\min}}{1 - x_{\min}^p} p x_i^{p-1} \mathbf{K}_i^0 \tag{6.46}$$

where $\mathbf{M}_i^0$ and $\mathbf{K}_i^0$ are mass and stiffness matrices of element i when it is solid.

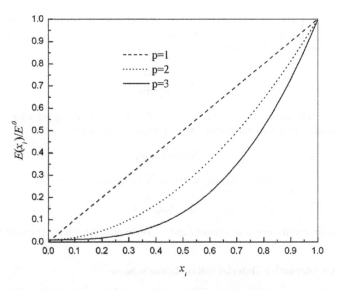

Figure 6.36 Alternative material interpolation scheme when $x_{\min} = 0.01$.

6.7.2 Sensitivity Number

From Equation (6.42), the sensitivity of the objective function, ω_i, can be expressed as

$$\frac{d\omega_j}{dx_i} = \frac{1}{2\omega_j \mathbf{u}_j^T \mathbf{M} \mathbf{u}_j} \left[2\frac{\partial \mathbf{u}_j^T}{\partial x_i} \left(\mathbf{K} - \omega_j^2 \mathbf{M} \right) \mathbf{u}_j + \mathbf{u}_j^T \left(\frac{\partial \mathbf{K}}{\partial x_i} - \omega_j^2 \frac{\partial \mathbf{M}}{\partial x_i} \right) \mathbf{u}_j \right] \tag{6.47}$$

Using equation (6.41), the above equation can be simplified as

$$\frac{d\omega_i}{dx_i} = \frac{1}{2\omega_i \mathbf{u}_i^T \mathbf{M} \mathbf{u}_i} \left[\mathbf{u}_i^T \left(\frac{\partial \mathbf{K}}{\partial x_i} - \omega_i^2 \frac{\partial \mathbf{M}}{\partial x_i} \right) \mathbf{u}_i \right] \tag{6.48}$$

Substituting Equation (6.46) into the above equation and assuming that the eigenvector $\mathbf{u}_j$ is normalized with respect to the global mass matrix $\mathbf{M}$, the sensitivity of the jth natural frequency can be found as

$$\frac{d\omega_j}{dx_i} = \frac{1}{2\omega_j} \mathbf{u}_j^T \left(\frac{1 - x_{\min}}{1 - x_{\min}^p} p x_i^{p-1} \mathbf{K}_i^0 - \omega_j^2 \mathbf{M}_i^0 \right) \mathbf{u}_j \tag{6.49}$$

The sensitivity numbers for solid and soft elements can be expressed explicitly as

$$\alpha_i = \frac{1}{p} \frac{d\omega_j}{dx_i} = \begin{cases} \dfrac{1}{2\omega_j} \mathbf{u}_j^T \left(\dfrac{1 - x_{\min}}{1 - x_{\min}^p} \mathbf{K}_i^0 - \dfrac{\omega_j^2}{p} \mathbf{M}_i^0 \right) \mathbf{u}_j & x_i = 1 \\[4mm] \dfrac{1}{2\omega_j} \mathbf{u}_j^T \left(\dfrac{x_{\min}^{p-1} - x_{\min}^p}{1 - x_{\min}^p} \mathbf{K}_i^0 - \dfrac{\omega_j^2}{p} \mathbf{M}_i^0 \right) \mathbf{u}_j & x_i = x_{\min} \end{cases} \tag{6.50}$$

When $x_{\min}$ tends to 0 (and $p > 1$), the sensitivity numbers can be simplified as

$$\alpha_i = \frac{1}{p}\frac{d\omega_j}{dx_i} = \begin{cases} \dfrac{1}{2\omega_j}\mathbf{u}_j^T\left(\mathbf{K}_i^0 - \dfrac{\omega_j^2}{p}\mathbf{M}_i^0\right)\mathbf{u}_j & x_i = 1 \\[2ex] -\dfrac{\omega_j}{2p}\mathbf{u}_j^T\mathbf{M}_i^0\mathbf{u}_j & x_i = x_{\min} \end{cases} \tag{6.51}$$

It should be noted that we still use $x_i = x_{\min}$ for void elements rather than $x_i = 0$ because the components of the eigenvector $\mathbf{u}_j$ related to void elements would not be found if these elements were totally eliminated from the design domain. Therefore, it is difficult to establish a rigorous hard-kill BESO method using the present material interpolation scheme.

6.7.3   Examples

6.7.3.1   2D Beams

Beam with simply supported ends
In this example, the objective is to maximize the fundamental frequency of a beam-like 2D structure shown in Figure 6.37 for a prescribed volume fraction $V_f = 50\%$ (Du and Olhoff 2007). The beam is simply supported at both ends. The rectangular design domain of 8 m × 1 m is divided into 320 × 40 four node plane stress elements. Young's modulus $E = 10$ MPa, Poisson's ratio $\nu = 0.3$ and mass density $\rho = 1$ kg/m^3.

The soft-kill BESO starts from the full design and gradually reduces the total volume of the structure using the evolutionary volume ratio $ER = 2\%$ and the maximum addition ratio $AR_{\max} = 2\%$. The design variable x_i is set to be either 1 or $x_{\min} = 10^{-6}$ and the filter radius $r_{\min}$ is 0.075 m. The penalty exponent $p = 3.0$ is used in calculating the sensitivity numbers. Figure 6.38 shows evolution histories of the first three natural frequencies as well as the volume fraction. It is seen that the first natural frequency increases and the second and third natural frequencies decrease as the volume fraction decreases. At later stages, the first and second frequencies become close, which is referred to as a multimodal case. However, the sensitivity of multiple frequencies of an equal value is not unique because of the lack of differentiability properties of the subspace spanned by the eigenvectors associated with the multiple frequencies (Seyranian *et al.* 1994). In BESO, a simple but effective way to solve this problem is to take an average of the two sensitivities of the first and second frequencies when the distance between the frequencies becomes less than a prescribed limit (Yang *et al.* 1999). It is seen from Figure 6.38 that the volume fraction reaches its constraint value of 50% after about 35 iterations and the first frequency converges to a constant value after 53 iterations.

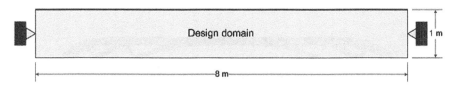

Figure 6.37   Design domain of a simply supported beam.

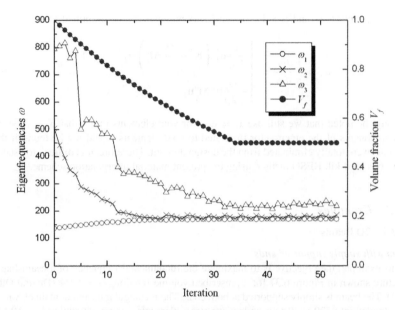

Figure 6.38 Evolution histories of the first three natural frequencies and the volume fraction.

The optimal design is given in Figure 6.39 with the fundamental frequency $\omega_1 = 171.5$ rad/s. The first two eigenmodes and the corresponding frequencies of the optimal design are given in Figure 6.40. It is noted that the above optimal topology, its natural frequencies and eigenmodes agree well with the results obtained by Du and Olhoff (2007) using an extended SIMP method. It has been demonstrated by Du and Olhoff (2007) that the first natural frequency of the optimal topology shown in Figure 6.39 is much higher than that produced by the homogenization method (Ma *et al.* 1995).

Beam with clamped ends
A beam of dimensions 14 cm $\times 2$ cm $\times 1$ cm is clamped on both sides as shown in Figure 6.41. Young's modulus $E = 100$ N/cm^2, Poisson's ratio $\nu = 0.3$ and mass density $\rho = 10^{-6}$ kg/cm^3 are assumed. A concentrated nonstructural mass $M = 1.4 \times 10^{-5}$ kg is placed at the centre. The rectangular design domain is divided into 280×40 four node plane stress elements. The BESO parameters are $ER = 2\%$, $AR_{\max} = 2\%$, $r_{\min} = 0.15$ cm and $p = 3.0$.

Figure 6.39 Optimal topology for maximizing the first frequency subject to a given volume.

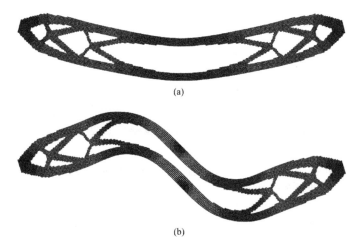

(a)

(b)

Figure 6.40 Eigenmodes of the optimal design: (a) the first eigenmode with $\omega_1 = 171.5$ rad/s; (b) the second eigenmode with $\omega_2 = 173.3$ rad/s.

First, we seek to find a solid-void design that maximizes the fundamental frequency subject to a volume fraction constraint of 50 % of the design domain. The void element are represented by an extremely small design variable $x_{min} = 10^{-6}$. Figure 6.42(a) shows the solid-void optimal design obtained using the soft-kill BESO method. The topology and the fundamental frequency both converge after 55 iterations. The fundamental frequency of the resultant topology is $\omega_1 = 33.7$ rad/s.

Next, we maximize the fundamental frequency of the clamped beam made of two different materials using the bi-material interpolation scheme (Bendsøe and Sigmund 1999; Huang and Xie 2009a). The stronger material is the same as that used above and the weaker material has properties $E = 20$ N/cm^2 and $\rho = 10^{-7}$ kg/cm^3. The volume fraction of the stronger material is set to be 50 % the whole design domain. No void is allowed in the design. Therefore, the weaker material will also occupy 50 % of the design domain. The final solution is obtained after 44 iterations and the resultant topology is shown in Figure 6.42(b), where dark and grey elements denote the stronger and weaker materials respectively. The fundamental frequency of the optimal design is 37.1rad/s.

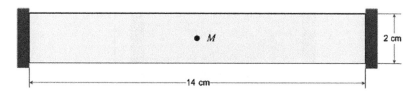

Figure 6.41 Design domain of a clamped beam with a concentrated mass.

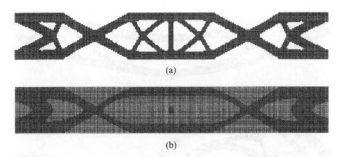

(a)

(b)

Figure 6.42 BESO solutions for the clamped beam: (a) solid-void optimal design with $\omega_1 = 33.7$ rad/s; (b) bi-material optimal design with $\omega_1 = 37.1$ rad/s.

6.7.3.2 3D Plates

Clamped square plate

Figure 6.43 shows a square plate of 20 m × 20 m with thickness $t = 1$ m. The plate is clamped at four edges. Young's modulus $E = 100$ GPa, Poisson's ratio $\nu = 0.3$ and mass density $\rho = 7800$ kg/m^3 are assumed. A concentrated nonstructural mass $M = 3.12 \times 10^5$ kg is attached to the centre of the plate. The design domain is modelled by 40 × 40 four node shell elements in ABAQUS (with each node having six degrees of freedom). The constraint on the volume fraction is set to be 50 % of the design domain. The BESO parameters $ER = 2\%$, $AR_{max}2\%$, $r_{min} = 1.5$ m, and $p = 3.0$ are used. The objective is to find the solid-void optimal design that maximizes a chosen natural frequency. The void elements are represented by a very small design variable $x_{min} = 10^{-6}$.

Figure 6.44(a) shows the optimal topology for maximizing the fundamental frequency while Figure 6.44(b) gives the evolution histories of the first three natural frequencies as well as the volume fraction. It is seen that although the material has been reduced by half, the

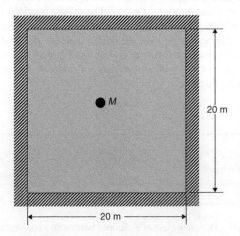

Figure 6.43 Design domain of a clamped square plate with a concentrated mass.

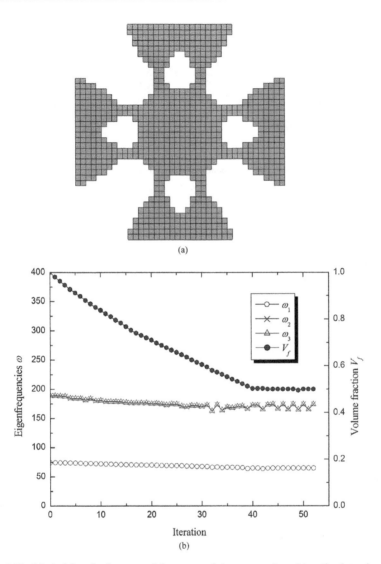

Figure 6.44 Maximizing the first natural frequency of the square plate: (a) optimal topology; (b) evolution histories of the first three natural frequencies and the volume fraction.

change in the fundamental frequency is relatively small. Initially, the first frequency gradually decreases as the total volume goes down. Once the volume constraint is satisfied, the volume is kept constant. After 52 iterations, the first frequency ω_1 converges to a constant value of 64.8rad/s. Figure 6.44(b) also reveals that the second and third eigenmodes are bimodal (i.e. their frequencies are identical or very close). For the final optimal design, ω_2 and ω_3 are both equal to 166.7rad/s.

(a)   (b)

Figure 6.45 The second and third eigenmodes of the initial plate corresponding to the bimodal natural frequencies $\omega_2 = \omega_3 = 189.0$ rad/s.

The second natural frequency of the initial design also corresponds to a bimodal mode as shown in Figure 6.45. In order to maximize the second natural frequency, the sensitivity number is calculated by taking the average of the sensitivity numbers for the related multiple frequencies (in this case, the second and third frequencies). The corresponding optimal design and evolution histories of the natural frequencies are shown in Figure 6.46. The frequencies of the optimal design are $\omega_1 = 59.5$ rad/s and $\omega_2 = \omega_3 = 192.1$ rad/s.

Simply supported rectangular plate
In this example, we consider the topology optimization of a rectangular plate which is simply supported at four edges (Tenek and Hagiwara 1994; Pedersen 2000). The dimensions of the plate are $L_1 = 0.15$ m, $L_2 = 0.1$ m and thickness $t = 0.00122$ m. Young's modulus $E = 70$ GPa, Poisson's ratio $v = 0.3$ and mass density $\rho = 2700$ kg/m^3 are assumed. A nonstructural mass $M = 0.003$ kg is attached to the centre of the plate. The design objective is to maximize the fundamental frequency. A constraint on the total mass is set to be 0.0277 kg (including the nonstructural mass).

Two types of designs are sought: (a) a solid-void design; (b) a bi-material design. In both cases, the BESO parameters used are $ER = 1\%$, $AR_{max} = 2\%$, $r_{min} = 0.006$ m and $p = 3.0$. For the solid-void design, the void elements are represented by $x_{min} = 10^{-6}$. For the bi-material design, Young's modulus and mass density of the weaker material are assumed to be $0.1E$ and 0.1ρ respectively. The resultant optimal designs are shown in Figure 6.47. It takes 66 and 73 iterations to obtain the convergent solid-void and bi-material designs, respectively. The final fundamental frequencies are $\omega_1 = 2162$ rad/s for the solid-void design and $\omega_1 = 2267$ rad/s for the bi-material design. It is noted that the bi-material design has a higher fundamental frequency than that of the solid-void design although the two structures have the same weight.

6.7.3.3 A 3D continuum structure

The present BESO method can be used for frequency optimization of 3D structures. Figure 6.48(a) shows a block of material simply supported at the four lower corners and a concentrated mass $M = 5000$ kg is attached to the centre of the bottom surface. The design domain is divided into $40 \times 40 \times 20$ eight node cubic elements. Young's modulus $E = 200$ GPa, Poisson's

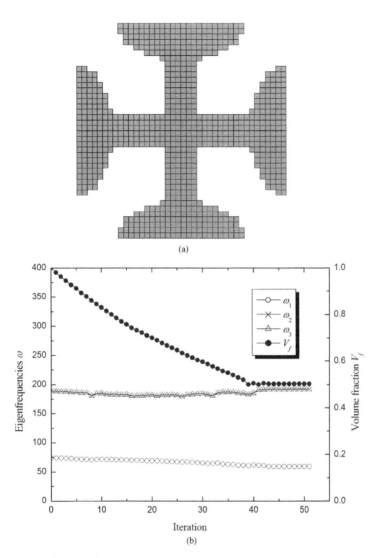

(a)

(b)

Figure 6.46 Maximizing the second natural frequency of the square plate: (a) optimal design; (b) evolution histories of the first three natural l frequencies and the volume fraction.

ratio $v = 0.3$ and material density $\rho = 7800 \text{ kg/m}^3$ are assumed. The design objective is to maximize the fundamental natural frequency of the structure subject to a volume fraction constraint of 15 % of the design domain.

In order to obtain a solid-void design, the design variables are restricted to be either 1 or $x_{\min} = 10^{-6}$. Other parameters used in the soft-kill BESO are $ER = 2\%$, $AR_{\max} = 2\%$,

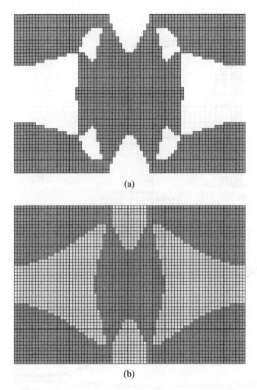

(a)

(b)

Figure 6.47 BESO solutions for a simply-supported rectangular plate: (a) solid-void optimal design with $\omega_1 = 2162$ rad/s; (b) bi-material optimal design with $\omega_1 = 2267$ rad/s.

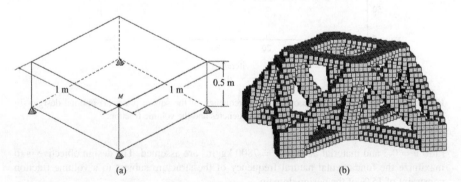

Figure 6.48 Frequency optimization of a 3D continuum structure with a concentrated mass: (a) design domain and boundary conditions; (b) optimal design with $\omega_1 = 502.6$ rad/s.

$p = 3.0$, and $r_{\min} = 0.075$ m. BESO starts from the initial full design and gradually reduces the total volume of the structure. The final optimal design shown in Figure 6.48(b) has a fundamental frequency of 502.6 rad/s.

6.8   Topology Optimization for Multiple Load Cases

Most real structures are subjected to different loads at different times. This is referred to as multiple load cases. Similarly, when a structure such as a bridge is subjected to a moving load, the force acting on the structure changes from one location to another. A moving load can be conveniently approximated to multiple load cases, by applying this load sequentially to a finite number of locations along the path of the moving load. With multiple load cases, the structure has to be designed to account for all load cases.

6.8.1   Sensitivity Number

The extension of the BESO method to structures with multiple load cases is straightforward. For example, the optimization problem can be formulated as one of minimizing a weighted average of the mean compliances of all load cases. Therefore, the topology optimization problem for multiple load cases can be stated as

$$\text{Minimize } f(x) = \sum_{k=1}^{M} w_k C_k$$

$$\text{Subject to: } V^* - \sum_{i=1}^{N} V_i x_i = 0 \qquad (6.52)$$

$$x_i = x_{\min} \text{ or } 1$$

where M is the total number of load cases, w_k is the prescribed weighting factor for the kth load case, C_k is the mean compliance of the kth load case, and $\sum_{k=1}^{M} w_k = 1$.

As the displacement field of one load case is independent of that of another load case, the sensitivity of the weighted objective function can be found as

$$\frac{df}{dx_i} = -\frac{1}{2} p x_i^{p-1} \sum_{k=1}^{M} w_k \left(\mathbf{u}_i^T \mathbf{K}_i^0 \mathbf{u}_i \right)_k \qquad (6.53)$$

Thus, the sensitivity number used in BESO can be defined as

$$\alpha_i = -\frac{1}{p} \frac{df}{dx_i} = \begin{cases} \dfrac{1}{2} \sum_{k=1}^{M} w_k (\mathbf{u}_i^T \mathbf{K}_i^0 \mathbf{u}_i)_k & x_i = 1 \\[4mm] \dfrac{1}{2} x_{\min}^{p-1} \sum_{k=1}^{M} w_k (\mathbf{u}_i^T \mathbf{K}_i^0 \mathbf{u}_i)_k & x_i = x_{\min} \end{cases} \qquad (6.54)$$

When p tends to infinity, the sensitivity number for the hard-kill BESO method can be simplified as

$$
\alpha_i = \begin{cases} \dfrac{1}{2} \displaystyle\sum_{k=1}^{M} w_k (\mathbf{u}_i^T \mathbf{K}_i^0 \mathbf{u}_i)_k & x_i = 1 \\[2ex] 0 & x_i = 0 \end{cases}
\tag{6.55}
$$

6.8.2   Examples

To verify the BESO procedures for structures with multiple load cases, two examples in Bendsøe and Sigmund (2003) are tested below, one using the soft-kill BESO and the other using the hard-kill BESO.

6.8.2.1   Example 1

The rectangular design domain is shown in Figure 6.49(a), which is divided into 120×60 four node plane stress elements. Young's modulus $E = 1$ and Poisson's ratio $\nu = 0.3$ are assumed. A constraint on the volume fraction is set to be 30% of the design domain. The soft-kill BESO method with $ER = 2\%$, $AR_{max} = 50\%$, $p = 3.0$, $x_{min} = 0.001$ and $r_{min} = 3.0$ are used in the example.

Figure 6.49(b) shows the optimal design when the two 1N forces are applied at the same time (i.e. in one load case) whereas Figure 6.49(c) gives the optimal design when the two forces are applied at different times (i.e. in two loads). Here, we choose to use equal weighting factors, i.e. $w_1 = w_2 = 0.5$, for the two load cases. It is noted that the multiple load case optimization results in a triangulated frame which is highly stable, whereas the single load case optimization leads to a trapezoidal frame which is far less stable. Figure 6.50 shows evolution histories of the objective function and the volume fraction for the topology optimization with multiple load cases.

6.8.2.2   Example 2

The rectangular design domain shown in Figure 5.51(a) is divided into 120×40 four node plane stress elements. Young's modulus $E = 1$ and Poisson's ratio $\nu = 0.3$ are assumed. A constraint on the volume fraction is set to be 30 % of the design domain. The hard-kill BESO method with $ER = 2\%$, $AR_{max} = 50\%$ and $r_{min} = 3$ is used. Equal weighting factors are applied to multiple load cases.

Figure 6.51 (b) shows optimal design for all the forces applied in one load case whereas Figure 6.51(c) gives the optimal design for three load cases of a single force. Once again, the multiple load case design gives a more stable structure of a triangulated frames. Although the single load case design is theoretically optimal under the forces it is designed for, the structure may not be robust enough to cope with the situation where the actual loading deviates from the assumed conditions.

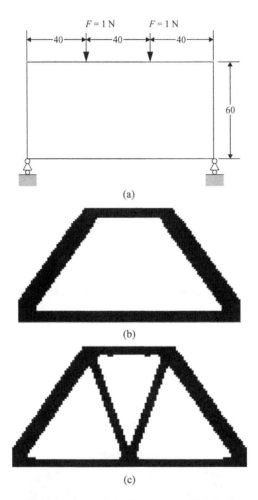

Figure 6.49 Topology optimization for single load case and multiple load cases: (a) design domain, loading and boundary conditions; (b) optimal design for one load case of two forces; (c) optimal topology for two load cases of a single force.

6.9   BESO Based on von Mises Stress

The original ESO method (Xie and Steven 1993) is based on von Mises stress, where lowly stressed material is assumed to be underutilized and therefore removed. Similarly, the removal and addition of material in the BESO method can also be decided by the stress level. Although the stress criterion is, to a large extent, equivalent to the stiffness criterion (Li *et al.* 1999), the problem statement for the topology optimization based on the von Mises stress differs significantly from that for stiffness optimization, as discussed below.

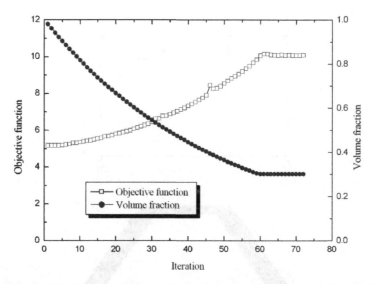

Figure 6.50 Evolution histories of objective function and volume fraction for optimization with multiple load cases.

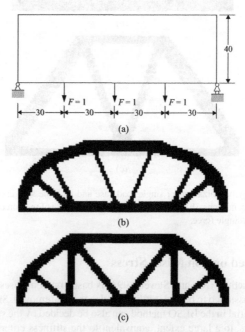

Figure 6.51 Topology optimization for single load case and multiple load cases: (a) design domain, loading and boundary conditions; (b) optimal design for all loads in one load case; (c) optimal topology for three load cases of a single force.

6.9.1   Sensitivity Number

Take a 2D isotropic elastic structure as an example. The von Mises stress of the ith element is defined as

$$\sigma_i^{vm} = \sqrt{\sigma_{xx}^2 + \sigma_{yy}^2 - \sigma_{xx}\sigma_{yy} + 3\tau_{xy}^2} = (\boldsymbol{\sigma}_i^T \mathbf{T}\boldsymbol{\sigma}_i)^{\frac{1}{2}} \tag{6.56}$$

where $\boldsymbol{\sigma}_i = \{\sigma_{xx}, \sigma_{yy}, \sigma_{xy}\}^T$ is the elemental stress vector, σ_{xx} and σ_{yy} are normal stresses in x and y directions, respectively, τ_{xy} is the shear stress, and $\mathbf{T}$ is the coefficient matrix given below

$$\mathbf{T} = \begin{bmatrix} 1 & -0.5 & 0 \\ -0.5 & 1 & 0 \\ 0 & 0 & 3 \end{bmatrix} \tag{6.57}$$

In finite element analysis, σ_i can be calculated from the nodal displacement vector $\mathbf{u}_i$ of element i as follows

$$\boldsymbol{\sigma}_i = \mathbf{DBu}_i \tag{6.58}$$

where $\mathbf{D}$ and $\mathbf{B}$ are the conventional elasticity and strain matrices respectively. When the power law material interpolation scheme Equation (4.2) is applied, the stress vector can be written as

$$\boldsymbol{\sigma}_i = x_{min}^P \mathbf{D}_0 \mathbf{Bu}_i \tag{6.59}$$

where $\mathbf{D}_0$ is the elasticity matrix of the solid element. Substituting the above equation into Equation (6.56) yields

$$\sigma_i^{vm} = x_{min}^P \left(\mathbf{u}_i^T \mathbf{B}^T \mathbf{D}_0^T \mathbf{T} \mathbf{D}_0 \mathbf{Bu}_i\right)^{\frac{1}{2}} = x_{min}^P \left(\boldsymbol{\sigma}_{i0}^T \mathbf{T}\boldsymbol{\sigma}_{i0}\right)^{\frac{1}{2}} = x_{min}^P \sigma_{i0}^{vm} \tag{6.60}$$

where σ_{i0}^{vm} is the von Mises stress of the solid element.

Obviously, when the design variable x_i changes, σ_i^{vm} varies. The following sensitivity number can be defined

$$\alpha_i = \frac{\sigma_i^{vm}}{x_i} = x_i^{P-1} \sigma_{i0}^{vm} \tag{6.61}$$

Therefore, the sensitivity numbers for solid and soft elements become

$$\alpha_i = \begin{cases} \sigma_{i0}^{vm} & x_i = 1 \\ x_{min}^{P-1} \sigma_{i0}^{vm} & x_i = x_{min} \end{cases} \tag{6.62}$$

When the penalty exponent tends to infinity, the sensitivity numbers reduce to

$$\alpha_i = \begin{cases} \sigma_{i0}^{vm} & x_i = 1 \\ 0 & x_i = 0 \end{cases} \tag{6.63}$$

where $x_i = x_{min}$ is replaced by $x_i = 0$ because a soft element with extremely small stiffness is equivalent to a void element.

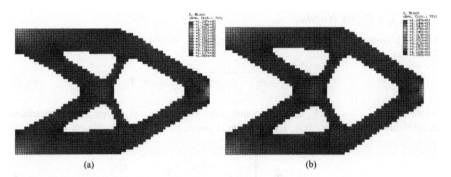

Figure 6.52 Topology optimization based on von Mises stress: (a) optimal design from soft-kill BESO method with $p = 3.0$; (b) optimal design from hard-kill BESO method.

6.9.2   Examples

6.9.2.1   A Short Cantilever

Both hard-kill and soft-kill BESO methods are applied to the cantilever structure shown in Figure 3.5, which is divided into 80×50 four node plane stress elements. The BESO parameters are $ER = 2\%$, $AR_{max} = 50\%$, $r_{min} = 3$ mm and $\tau = 0.1\%$. For the soft-kill BESO, the lower bound of the material density x_{min} and the penalty exponent p are set to be 0.001 and 3.0, respectively.

Figure 6.52 shows the final optimal topologies from the soft-kill and hard-kill BESO methods after 56 and 43 iterations respectively. Their corresponding mean compliances are 1.895 Nmm and 1.870 Nmm. Both the topologies and the mean compliances are very close to the BESO results for stiffness optimization given in Chapter 3. It confirms the previous observation by Li *et al.* (1999) that the stress criterion is equivalent to the stiffness criterion.

6.9.2.2   Michell Type Structures

In this example, we apply the above hard-kill BESO procedure to Michell type structures (Michell 1904; Hemp 1973). The design domain shown in Figure 2.3 is divided into 200×100 four node plane stress elements. The two corners at the bottom are simply supported. The same BESO parameters as those in the previous example are used except for r_{min} which is now set to be 0.15m. When the volume constraint is set to be 20% and 10% of the design domain, we obtain the optimal designs shown in Figures 6.53(a) and (b). When the right hand side support is changed to a roller as shown in Figure 2.5, the resulting optimal designs change to those shown in Figure 6.54.

The design domain of another Michell type structure is shown in Figure 6.55. An optimal structure is to be found to transfer a vertical force F to the circular fixed support. The radius of circular support is equal to 10. The structure is assumed to be under plane stress conditions and the rectangular domain (including the circular area of the support) is divided into 120×80 four node elements. The same BESO parameters as those specified in section 6.9.2.1 are used except for r_{min} which is now set to be 1. The resulting optimal designs for volume fractions of 50% and 30% are shown in Figures 6.56(a) and (b), respectively.

Figure 6.53 Optimal topologies for a Michell type structure with two simple supports: (a) $V_f = 0.2$ and (b) $V_f = 0.1$.

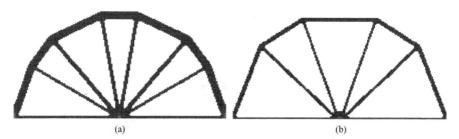

Figure 6.54 Optimal topologies for a Michell type structure with one simple support and one roller: (a) $V_f = 0.2$ and (b) $V_f = 0.1$.

Figure 6.55 Design domain, loading and boundary conditions for a Michell type structure.

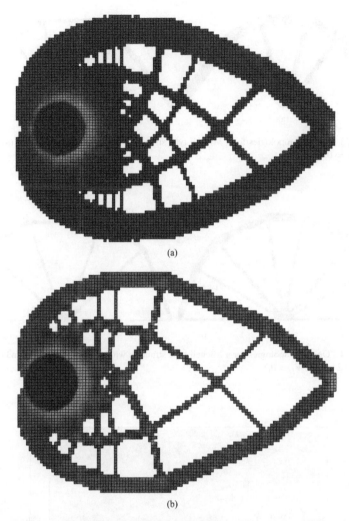

(a)

(b)

Figure 6.56   Optimal designs for a Michell type structure: (a) $V_f = 0.5$ and (b) $V_f = 0.3$.

It is noted that the BESO topologies for Michell type structures are very close to the original Michell trusses (Hemp 1973; Michell 1904). Compared with the original ESO procedure, the present BESO with mesh-independency filter produces much better solutions without checkerboard patterns. It should be pointed out that a volume constraint cannot be adequately imposed in the original ESO procedure because it is aimed at reaching a steady state of the stress level rather than moving towards a specific structural volume. That is not a proper way to solve the optimization problem with a volume constraint. The present BESO method can produce designs that satisfy the volume constraint exactly.

6.10 Conclusion

This chapter has extended the BESO method to a wide range of topology optimization problems. Except for Section 6.9 for von Mises stress, the extensions of the BESO method are based on the sensitivity analysis of the objective functions and constraints. The sensitivity numbers for discrete design variables are established utilizing various material interpolation schemes. Combined with optimality criteria, the extended BESO algorithms ensure the convergent solution to be an optimum, or at least a local optimum. The extended BESO methods inherit the advantages of the basic BESO algorithm such as simplicity, high efficiency and effectiveness. Usually the hard-kill BESO method is computationally more efficient than the soft-kill BESO method. However, as we have seen in this chapter, the effectiveness of the hard-kill approach depends on the optimization problem. When a new type of topology optimization problem is encountered, it is suggested that one should develop a soft-kill BESO method first and then explore the possibility of establishing a corresponding hard-kill approach.

References

Ansola, R., Canales, J. and Tarrago, J.A. (2006). An efficient sensitivity computation strategy for the evolutionary structural optimization (ESO) of continuum structures subjected to self-weight loads. *Finite Elements in Analysis and Design* **42**: 1220–30.

Bendsøe, M.P. and Sigmund, O. (1999). Material interpolation schemes in topology optimization. *Archive of Appl. Mech.* **69**: 635–54.

Bendsøe, M.P. and Sigmund, O. (2003). *Topology Optimization: Theory, Method and Application*. Berlin: Springer.

Bruyneel, M. and Duysinx, P. (2005). Note on topology optimization of continuum structures including self-weight. *Struct. Multidisc. Optim.* **29**: 245–56.

Chiandussi, G. (2006). On the solution of a minimum compliance topology optimization problem by optimality criteria without a priori volume constraint specification. *Comput. Mech.* **38**(1): 77–99.

Du, J. and Olhoff, N. (2007). Topological design of freely vibrating continuum structures for maximum values of simple and multiple eigenfrequencies and frequency gaps. *Struct. Multidisc. Optim.* **34**: 91–110.

Hemp, W.S. (1973). Michell's structrual continua. In: *Optimum Structures*. Chap. 4, Oxford: Clarendon Press.

Huang, X. and Xie, Y.M. (2007). Convergent and mesh-independent solutions for bi-directional evolutionary structural optimization method. *Finite Elements in Analysis and Design* **43**(14): 1039–49.

Huang, X. and Xie, Y.M. (2008). Optimal design of periodic structures using evolutionary topology optimization. *Struct. Multidisc. Optim.* **36**(6): 597–609.

Huang, X. and Xie, Y.M. (2009a). Bi-directional evolutionary topology optimization of continuum structures with one or multiple materials. *Comput. Mech.* **43**: 393–401.

Huang, X. and Xie, Y.M. (2009b). Evaluation of topology optimization procedures for minimizing material volume subject to a given compliance. Submitted to *Advances in Structural Engineering*.

Huang, X. and Xie, Y.M. (2009c). Evolutionary topology optimization of continuum structures including self-weight loads. Submitted to *Finite Element in Analysis and Design*.

Huang, X. and Xie, Y.M. (2009d). Evolutionary topology optimization of continuum structures with an additional displacement constraint. *Struct. Multidisc. Optim.* DOI 10.1007/s00158-009-0382-4

Huang, X. and Xie, Y.M. (2009e). A further review of ESO type methods for topology optimization. *Struct. Multidisc. Optim.* (to appear).

Huang, X., Zuo, Z.H. and Xie, Y.M. (2009). Evolutionary topology optimization of vibrating continuum structures for natural frequencies. *Comput. & Struct.* DOI 10.1016/j.compstruc.2009.11.011

Kočvara, M. (1997). Topology optimization with displacement constraints: a bilevel programming approach. *Struct. Optim.* **14**: 256–63.

Kosaka, I. and Swan, C.C. (1999). A symmetry reduction method for continuum structural topology optimization. *Comput. Struct.* **70**: 47–61.

Li, Q., Steven, G.P. and Xie, Y.M. (1999). On equivalence between stress criterion and stiffness criterion in evolutionary structural optimization. *Struct. Multidisc. Optim.* **18**(1): 67–73.

Ma, Z.D., Cheng, H.C. and Kikuchi N. (1995). Topological design for vibrating structures. *Comput. Meth. Appl. Mech. Eng.* **121**: 259–80.

Michell, A.G.M. (1904). The limits of economy of material in frame-structures. *Phil. Mag.* **8**: 589–97.

Pedersen, N.L. (2000). Maximization of eigenvalues using topology optimization. *Struct Multidisc Optim.* **20**: 2–11.

Rozvany, G.I.N. (2009). A critical review of established methods of structrual topology optimization. *Struct. Multidisc. Optim.* **37**(3): 217–37.

Seyranian, A.P., Lund, E. and Olhoff, N. (1994). Multiple eigenvalues in structural optimization problems. *Struct. Optim.* **8**(4): 207–27.

Sigmund, O. (1994). Materials with prescribed constitutive parameters: an inverse homogenization problem. *Int. J. Solids and Struct.* **31**(17): 2313–29.

Sigmund, O. and Torquato, S. (1997). Design of materials with extreme thermal expansion using a three-phase topology optimization method. *J. Mech. Phys. Solids* **45**: 1037–67.

Stolpe, M. and Svanberg, K. (2001). An alternative interpolation model for minimum compliance topology optimizaiton. *Struct. Multidisc. Optim.* **22**: 116–24.

Tenek, L.H. and Hagiwara, I. (1994). Eigenfrequency maximization of plates by optimization of topology using homogenization and mathematical programming. *JSME Int. J.* **37**: 667–77.

Wadley, H.N.G., Fleck, N.A. and Evan, A.G. (2003). Fabrication and structural performance of periodic cellular metal sandwich structures. *Composites Science and Technology* **63**: 2331–43.

Xie, Y.M. and Steven, G.P. (1993). A simple evolutionary procedure for structural optimization. *Computers & Structures* **49**: 885–96.

Xie, Y.M. and Steven, G.P. (1996). Evolutionary structural optimization for dynamic problems. *Comput. Struct.* **58**: 1067–73.

Yang, X.Y., Xie, Y.M. and Steven, G.P. (2005). Evolutionary methods for topology optimization of continuous structures with design dependent loads. *Comput. Struct.* **83**: 956–63.

Yang, X.Y., Xie, Y.M., Steven, G.P. and Querin, O.M. (1999). Topology optimization for frequencies using an evolutionary method. *J. Struct. Engng* **125**(12): 1432–8.

Zhang, W. and Sun, S. (2006). Scale-related topology optimization of cellular materials and structures. *Int. J. Numer. Meth. Engng.* **68**: 993–1011.

7

Topology Optimization of Nonlinear Continuum Structures

7.1 Introduction

Most works on structural optimization are concerned with the optimal design of structures with linear material and small deformation. Extending the optimization methods to applications involving nonlinear material and large deformation is not straightforward. Using sensitivity/gradient based optimization methods, a number of researchers have considered topology optimization of geometrically nonlinear structures (Buhl *et al.* 2000; Gea and Luo 2001; Pedersen *et al.* 2001; Bruns and Tortorelli 2003). Topology optimization of structures with nonlinear materials has also been conducted by several researchers (Yuge and Kikuchi 1995; Bendsøe *et al.* 1996; Pedersen 1998). However, there has been very limited research on topology optimization with both geometrical and material nonlinearities, one exception being the work of Jung and Gea (2004) who used a generalized convex approximation method.

Topology optimization of nonlinear structures requires a much larger amount of computational time than that of linear structures. Thus, computational efficiency of the optimization method is critical, especially for 3D structures. Moreover, in optimization methods, such as SIMP, intermediate material densities appear in the topology results. As the topology develops, large displacements may cause the tangential stiffness matrix of low-density elements to become indefinite or even negatively definite (Buhl *et al.* 2000; Bruns and Tortorelli 2003). Therefore some additional schemes, such as totally removing low-density elements (Bruns and Tortorelli 2003) or relaxing the convergence criterion (Buhl *et al.* 2000), must be devised in order to circumvent the numerical problems.

The hard-kill BESO method is capable of overcoming these problems. In the hard-kill BESO method, soft elements are totally removed from the design domain and therefore there is no convergence problem caused by low-density elements. On the other hand, the removal of elements reduces the size of the finite element model and improves the computational efficiency of the optimization process. In this chapter, we present a hard-kill BESO method for topology optimization of nonlinear continuum structures (Huang and Xie 2007; 2008).

Evolutionary Topology Optimization of Continuum Structures: Methods and Applications   Xiaodong Huang and Mike Xie
© 2010 John Wiley & Sons, Ltd

7.2   Objective Functions and Nonlinear Analysis

For many industrial applications, the maximum stiffness of a structure is pursued. Consider a nonlinear structure subjected to an applied load which increases monotonously with displacement up to a maximum value, **F**. The corresponding nonlinear force-displacement curve is depicted in Figure 7.1(a). To maximize the structural stiffness, the natural choice of the optimization objective is to minimize the displacement $\mathbf{U}^*$ or the end compliance $\mathbf{F}^T \mathbf{U}^*$ in the deflected configuration (Buhl *et al.* 2000). However, minimization of the end compliance may result in degenerated structures which can only support the maximum load they are designed for. To avoid this problem and make sure that the structure is stable for any load up to the maximum design load, one may minimize the complementary work W^C shown as the shaded area in Figure 7.1(a). Thus, the optimization problem for maximizing stiffness with a volume constraint can be formulated using the elements as the design variables as

$$
\begin{aligned}
\text{Minimize } f_1(x) = W^C &= \lim_{n \to \infty} \left[\frac{1}{2} \sum_{i=1}^{n} \Delta \mathbf{F}^T \left(\mathbf{U}_i + \mathbf{U}_{i-1} \right) \right] \\
\text{Subject to : } g = V^* &- \sum_{e=1}^{M} V_e x_e = 0 \\
x_e &= x_{\min} \text{ or } 1
\end{aligned}
\tag{7.1}
$$

where $\mathbf{U}$ is the displacement vector, i is the incremental number of the load vector and n is the total number of load increments. The size of the increment is determined by $\Delta \mathbf{F} = \mathbf{F}/n$. V_e is the volume of an individual element and V^* is the prescribed objective (target) volume for the final design. The binary design variable x_e denotes the soft ($x_{\min}$) or solid (1) element. M is the total element number in the finite element model.

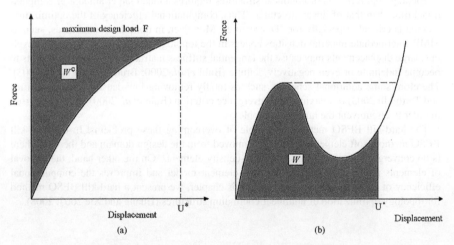

Figure 7.1   Typical load-deflection curves in nonlinear finite element analysis: (a) objective function W^C for force control; (b) objective function W for displacement control.

Consider another typical nonlinear load-displacement curve of the applied force against displacement depicted in Figure 7.1(b). If the structure is physically loaded under force control, the displacement will snap through along the dotted line. Alternatively, under displacement control the trajectory can follow the whole equilibrium path. For most applications, displacement control is more practical. For example, one could use displacement control to study the crushing distance of energy absorbing structures. To maximize the structural stiffness (or the carried-load) at any displacement from 0 to the design displacement $\mathbf{U}^*$, one may maximize the total external work W shown as the shaded area in Figure 7.1(b): which is equal to the total strain energy under quasi-static conditions. Thus, the optimization problem with a volume constraint can be formulated with as

$$
\begin{aligned}
\text{Maximize } f_2(x) = W &= \lim_{n\to\infty} \left[\frac{1}{2} \sum_{i=1}^{n} (\mathbf{U}_i^T - \mathbf{U}_{i-1}^T)(\mathbf{F}_i + \mathbf{F}_{i-1}) \right] \\
\text{Subject to}: \ g &= V^* - \sum_{e=1}^{M} V_e x_e = 0 \\
x_e &= x_{\min} \text{ or } 1
\end{aligned}
\tag{7.2}
$$

where the final displacement vector $\mathbf{U}_n = \mathbf{U}^*$ and the size of the increment is determined by $\Delta\mathbf{U} = \mathbf{U}^*/n$.

While the response of a linear structure can be determined by simply solving a set of linear equations, the equilibrium of a nonlinear structure needs to be found by an iterative procedure. The residual force, $\mathbf{R}$, is defined as the discrepancy between the internal force vector and the external force vector as

$$
\mathbf{R} = \mathbf{F} - \mathbf{F}^{\text{int}} = 0
\tag{7.3}
$$

The internal force vector can be expressed as

$$
\mathbf{F}^{\text{int}} = \sum_{e=1}^{M} \int_e \mathbf{C}^{\mathbf{e}T} \mathbf{B} \sigma \, dv = \sum_{e=1}^{M} \mathbf{C}^{\mathbf{e}T} \mathbf{F}_e^{\text{int}}
\tag{7.4}
$$

where $\mathbf{F}_e^{\text{int}}$ is the nodal force vector of an element and $\mathbf{C}^{\mathbf{e}}$ is a matrix which transforms the nodal force vector of an element to the global nodal force vector.

Quite often, the equilibrium Equation (7.3) is solved incrementally and iteratively using the Newton-Raphson method, which requires the determination of the tangential stiffness matrix in each step as

$$
\mathbf{K}^{\mathbf{t}} = -\frac{\partial\mathbf{R}}{\partial\mathbf{U}}
\tag{7.5}
$$

Sometimes, more sophisticated algorithms than the Newton-Raphson method need to be employed, especially for highly nonlinear problems. Details of nonlinear finite element analysis techniques can be found in textbooks (e.g. Crisfield 1991).

7.3 Sensitivity Analysis and Sensitivity Number for Force Control

Consider a nonlinear structural system with design independent loads. Assuming that the design variable varies continuously from 1 to 0, the sensitivity of the complementary work with respect to the design variable x_e is

$$\frac{\partial f_1(x)}{\partial x_e} = \lim_{n \to \infty} \left[\frac{1}{2} \sum_{i=1}^{n} (\mathbf{F}_i^T - \mathbf{F}_{i-1}^T) \left(\frac{\partial \mathbf{U}_i}{\partial x_e} + \frac{\partial \mathbf{U}_{i-1}}{\partial x_e} \right) \right] \qquad (7.6)$$

An adjoint equation is introduced by adding a series of vectors of Lagrangian multipliers λ_i into the objective function (Buhl *et al.* 2000) as

$$f_1(x) = \lim_{n \to \infty} \frac{1}{2} \sum_{i=1}^{n} \left[(\mathbf{F}_i^T - \mathbf{F}_{i-1}^T)(\mathbf{U}_i + \mathbf{U}_{i-1}) + \lambda_i^T (\mathbf{R}_i + \mathbf{R}_{i-1}) \right] \qquad (7.7)$$

where $\mathbf{R}_i$ and $\mathbf{R}_{i-1}$ are the residual forces in load increment steps i and $i - 1$. Thus

$$\mathbf{R}_i + \mathbf{R}_{i-1} = \mathbf{F}_i - \mathbf{F}_i^{\text{int}} + \mathbf{F}_{i-1} - \mathbf{F}_{i-1}^{\text{int}} = 0 \qquad (7.8)$$

Because both $\mathbf{R}_i$ and $\mathbf{R}_{i-1}$ are equal to zero, the modified objective function in (7.7) is the same as the original objective function in (7.1). The sensitivity of the modified objective function is

$$\frac{\partial f_1(x)}{\partial x_e} = \lim_{n \to \infty} \frac{1}{2} \sum_{i=1}^{n} \left[(\mathbf{F}_i^T - \mathbf{F}_{i-1}^T) \left(\frac{\partial \mathbf{U}_i}{\partial x_e} + \frac{\partial \mathbf{U}_{i-1}}{\partial x_e} \right) \right.$$
$$\left. + \lambda_i^T \left(\frac{\partial \mathbf{R}_i}{\partial \mathbf{U}_i} \frac{\partial \mathbf{U}_i}{\partial x_e} + \frac{\partial \mathbf{R}_{i-1}}{\partial \mathbf{U}_{i-1}} \frac{\partial \mathbf{U}_{i-1}}{\partial x_e} + \frac{\partial (\mathbf{R}_i + \mathbf{R}_{i-1})}{\partial x_e} \right) \right] \qquad (7.9)$$

Note that the derivative of λ_i is not included in the above equation because it would be multiplied by $(\mathbf{R}_i + \mathbf{R}_{i-1})$ which is zero. It is assumed that there is a linear force-displacement relationship in a small load increment. Therefore $\frac{\partial \mathbf{R}_i}{\partial \mathbf{U}_i} = \frac{\partial \mathbf{R}_{i-1}}{\partial \mathbf{U}_{i-1}} = -\mathbf{K}_i^t$, where $\mathbf{K}_i^t$ is the tangential stiffness in the ith step. By substituting this relationship into Equation (7.9), the sensitivity of the modified objective function can be rewritten as

$$\frac{\partial f_1(x)}{\partial x_e} = \lim_{n \to \infty} \frac{1}{2} \sum_{i=1}^{n} \left[(\mathbf{F}_i^T - \mathbf{F}_{i-1}^T - \lambda_i^T \mathbf{K}_i^t) \left(\frac{\partial \mathbf{U}_i}{\partial x_e} + \frac{\partial \mathbf{U}_{i-1}}{\partial x_e} \right) + \lambda_i^T \frac{\partial (\mathbf{R}_i + \mathbf{R}_{i-1})}{\partial x_e} \right] \qquad (7.10)$$

In order to eliminate the unknowns $\frac{\partial \mathbf{U}_i}{\partial x_e} + \frac{\partial \mathbf{U}_{i-1}}{\partial x_e}$, λ_i can be chosen as

$$\mathbf{K}_i^t \lambda_i = \mathbf{F}_i - \mathbf{F}_{i-1} \qquad (7.11)$$

This equation defines the adjoint system. From the assumption of linear force-displacement relationship in a small increment, the increment of the force can be approximately expressed by

$$\mathbf{K}_i^t (\mathbf{U}_i - \mathbf{U}_{i-1}) = \mathbf{F}_i - \mathbf{F}_{i-1} \qquad (7.12)$$

Comparing Equation (7.11) with Equation (7.12), we obtain λ_i as

$$\lambda_i = \mathbf{U}_i - \mathbf{U}_{i-1} \tag{7.13}$$

By substituting λ_i into Equation (7.10) and utilizing Equation (7.8), the sensitivity of the objective function is expressed by

$$\frac{\partial f_1(x)}{\partial x_e} = -\lim_{n \to \infty} \frac{1}{2} \sum_{i=1}^{n} \left(\mathbf{U}_i^T - \mathbf{U}_{i-1}^T \right) \left(\frac{\partial \mathbf{F}_i^{\text{int}}}{\partial x_e} + \frac{\partial \mathbf{F}_{i-1}^{\text{int}}}{\partial x_e} \right) \tag{7.14}$$

Material nonlinearity can be conveniently modelled by a general relationship between the effective stress and the effective strain as

$$\bar{\sigma} = K \Phi(\bar{\varepsilon}) \tag{7.15}$$

where function $\Phi(\bar{\varepsilon})$ is a general function representing the material characteristics and K is a constant reference modulus of elasticity. For example, in the power-law material model, $\Phi(\bar{\varepsilon})$ is in the form of $\bar{\varepsilon}^n$ where n is the work-hardening exponent. In order to consider both solid and void elements (as well as soft element), the following interpolation scheme is introduced

$$\bar{\sigma}(x_e) = x_e^p K \Phi(\bar{\varepsilon}^0) \tag{7.16}$$

where $\bar{\varepsilon}^0$ denotes the effective strain of solid material. As a result, the internal force vector can be expressed by

$$\mathbf{F}^{\text{int}} = \sum_{e=1}^{M} x_e^p \mathbf{C}^{\mathbf{eT}} \mathbf{F}_{e0}^{\text{int}} \tag{7.17}$$

where $\mathbf{F}_{e0}^{\text{int}}$ is the internal force vector for solid element. By substituting the above equation into Equation (7.14), we have

$$\frac{\partial f_1(x)}{\partial x_e} = -\lim_{n \to \infty} \frac{1}{2} p x_e^{p-1} \sum_{i=1}^{n} \left(\mathbf{U}_i^T - \mathbf{U}_{i-1}^T \right) \left(\mathbf{C}^{eT} \mathbf{F}_{e,i}^{\text{int}} + \mathbf{C}^{eT} \mathbf{F}_{e,i-1}^{\text{int}} \right) \tag{7.18}$$

where $\mathbf{F}_{e,i}^{\text{int}}$ denotes the internal force vector for solid element in the ith increment. The negative sign indicates that the complementary work decreases when the design variable x_e increases. In order to minimize the complementary work, we define the elemental sensitivity number, which denotes the relative ranking of the elemental sensitivity, as

$$\begin{aligned} \alpha^e &= -\frac{1}{p} \frac{\partial f_1(x)}{\partial x_e} = x_e^{p-1} \lim_{n \to \infty} \frac{1}{2} \sum_{i=1}^{n} \left(\mathbf{U}_i^T - \mathbf{U}_{i-1}^T \right) \left(\mathbf{C}^{eT} \mathbf{F}_{e,i}^{\text{int}} + \mathbf{C}^{eT} \mathbf{F}_{e,i-1}^{\text{int}} \right) \\ &= x_e^{p-1} \lim_{n \to \infty} \sum_{i=1}^{n} \left(E_i^e - E_{i-1}^e \right) = x_e^{p-1} E_n^e \end{aligned} \tag{7.19}$$

where E_n^e is the final elemental elastic and plastic strain energy.

In the BESO method only two discrete design variables $x_{\min}$ and 1 are used. Therefore, the sensitivity number can be expressed explicitly as

$$\alpha_e = \begin{cases} E_n^e & when \; x_e = 1 \\ x_{\min}^{p-1} E_n^e & when \; x_e = x_{\min} \end{cases} \tag{7.20}$$

However, as mentioned previously, large displacements in the minimum density elements may cause the tangential stiffness matrix to become indefinite or even negatively definite (Bruns and Tortorelli 2003). This will lead to convergence difficulty in finite element analysis. Obviously, the hard-kill BESO method is a natural way to circumvent the above problem. The sensitivity numbers for the hard-kill BESO method is obtained as p tends to infinity.

$$\alpha_e = \begin{cases} E_n^e & x_e = 1 \\ 0 & x_e = 0 \end{cases} \tag{7.21}$$

where $x_e = x_{\min}$ is replaced with $x_e = 0$ for the void element. The above sensitivity number for nonlinear structures would degenerate to the final elastic strain energy for linear structures. However, one should note the considerable differences in the objective function, the finite element analysis and the plastic component of the sensitivity number for nonlinear structures although Equation (7.21) appears to be identical to the sensitivity formula for linear structures discussed in Chapter 3. Nevertheless, one great advantage of the similarity between the formulations for linear and nonlinear structures is that very few changes need to be made to the BESO computer code in order to extend its application from linear to nonlinear structures.

7.4 Sensitivity Analysis and Sensitivity Number for Displacement Control

Consider a nonlinear structure under displacement loading. The displacement is increased step-by-step up to the prescribed displacement $\mathbf{U}^*$. The sensitivity of the objective function is

$$\frac{\partial f_2(x)}{\partial x_e} = \lim_{n \to \infty} \left[\frac{1}{2} \sum_{i=1}^{n} \left(\mathbf{U}_i^T - \mathbf{U}_{i-1}^T \right) \left(\frac{\partial \mathbf{F}_i}{\partial x_e} + \frac{\partial \mathbf{F}_{i-1}}{\partial x_e} \right) + \frac{1}{2} \sum_{i=1}^{n} \left(\frac{\partial \mathbf{U}_i^T}{\partial x} - \frac{\partial \mathbf{U}_{i-1}^T}{\partial x} \right) \left(\mathbf{F}_i + \mathbf{F}_{i-1} \right) \right] \tag{7.22}$$

It is noted that the second term is zero because the variation of displacement equals zero, i.e. $\frac{\partial \mathbf{U}_i^T}{\partial x_e} = 0$ and $\frac{\partial \mathbf{U}_{i-1}^T}{\partial x_e} = 0$, at the controlled freedoms (where the displacement loading is applied) and the external forces $\mathbf{F}_i$ and $\mathbf{F}_{i-1}$ equal zero at other freedoms. Similarly, an adjoint equation is introduced by adding a series of vectors of Lagrangian multipliers λ_i to the objective function. Thus

$$f_2(x) = \lim_{n \to \infty} \frac{1}{2} \sum_{i=1}^{n} \left[\left(\mathbf{U}_i^T - \mathbf{U}_{i-1}^T \right) \left(\mathbf{F}_i + \mathbf{F}_{i-1} \right) - \lambda_i^T \left(\mathbf{R}_i + \mathbf{R}_{i-1} \right) \right] \tag{7.23}$$

The modified objective function in (7.23) is the same as the original objective function in (7.2) because $\mathbf{R}_i + \mathbf{R}_{i-1} = 0$. From Equation (7.8), the sensitivity of the modified objective function becomes

$$\frac{\partial f_2(x)}{\partial x_e} = \lim_{n \to \infty} \frac{1}{2} \sum_{i=1}^{n} \left[\left(\mathbf{U}_i^T - \mathbf{U}_{i-1}^T \right) \left(\frac{\partial \mathbf{F}_i}{\partial x_e} + \frac{\partial \mathbf{F}_{i-1}}{\partial x_e} \right) - \lambda_i^T \left(\frac{\partial \mathbf{F}_i}{\partial x_e} - \frac{\partial \mathbf{F}_i^{\text{int}}}{\partial x_e} + \frac{\partial \mathbf{F}_{i-1}}{\partial x_e} - \frac{\partial \mathbf{F}_{i-1}^{\text{int}}}{\partial x_e} \right) \right]$$

(7.24)

To eliminate the unknowns $\frac{\partial \mathbf{F}_i}{\partial x_e} + \frac{\partial \mathbf{F}_{i-1}}{\partial x_e}$, let the adjoint variable λ_i be

$$\lambda_i = \mathbf{U}_i - \mathbf{U}_{i-1} \tag{7.25}$$

By substituting λ_i into Equation (7.24), the sensitivity of the objective function is expressed by

$$\frac{\partial f_2(x)}{\partial x_e} = \lim_{n \to \infty} \frac{1}{2} \sum_{i=1}^{n} \left(\mathbf{U}_i^T - \mathbf{U}_{i-1}^T \right) \left(\frac{\partial \mathbf{F}_i^{\text{int}}}{\partial x_e} + \frac{\partial \mathbf{F}_{i-1}^{\text{int}}}{\partial x_e} \right) \tag{7.26}$$

The above equation is the same as Equation (7.14) except for positive sign here. Equation (7.26) indicates that the total external work increases when the design variable x_e increases. With the material interpolation scheme equation (7.16), the elemental sensitivity number can be defined as

$$\alpha^e = \frac{1}{p} \frac{\partial f_2(x)}{\partial x_e} = x_e^{p-1} \lim_{n \to \infty} \sum_{i=1}^{n} \left(E_i^e - E_{i-1}^e \right) = x_e^{p-1} E_n^e \tag{7.27}$$

The above equation shows that the same expression of the sensitivity number can be used for nonlinear structures with both force and displacement controls. However, for the displacement loading we seek to maximize the external work of the design, while when a structure is loaded under force control the objective is to minimize the complementary work.

7.5 BESO Procedure for Nonlinear Structures

The BESO procedure for nonlinear structures is similar to that for linear structures except that nonlinear finite element analysis must be used. The evolutionary optimization procedure for nonlinear structures is given as follows:

1. Discretize the design domain using a finite element mesh for the given boundary and loading conditions. Assign the initial property values (0 or 1) to the elements to construct an initial design.
2. Perform nonlinear finite element analysis on the current design and then calculate the elemental sensitivity numbers.
3. Filter the elemental sensitivity numbers using the mesh-independency filter.
4. Average the elemental sensitivity numbers using their historical information and then save the resulting sensitivity numbers for the next iteration.
5. Determine the target volume for the next iteration.
6. Reset the property values of elements. For solid elements (1), the property value is switched from 1 to 0 if its sensitivity number is smaller than the threshold. For void elements (0), the

property value is switched from 0 to 1 if its sensitivity number is larger than the threshold. Then construct a new design using elements with property value 1 for the next FE analysis.
7. Check the boundary and loading conditions of the new design.
8. Repeat steps 2–7 until the objective volume (V^*) is achieved and the complementary work for force control or the external work for displacement control has converged.

7.6 Examples of Nonlinear Structures under Force Control

7.6.1 Geometrically Nonlinear Structure

The first example considers the stiffness optimization of a slender cantilever (Buhl *et al.* 2000) under a concentrated loading as shown in Figure 7.2. The cantilever is 1 m in length, 0.25 m in width and 0.1 m in thickness. It is fixed at one end and free at the other. It is assumed that the available material can only cover 50 % of the design domain. The material is linear elastic with Young's modulus $E = 3$ GPa and Poisson's ratio $v = 0.4$. The BESO parameters used in this example are $ER = 1\,\%$, $AR_{max} = 1\,\%$, filter radius $r = 0.02$ m and $\tau = 1\,\%$.

If the optimization problem is solved using linear finite element analysis, the optimal topology would not depend on the magnitude of the applied load. Topology optimization using linear finite element analysis is first carried out to find the linear design. As expected the topology of the linear design is symmetric, as shown in Figure 7.3(a). The nonlinear designs for $F = 60$ kN and 144 kN obtained from the BESO method using the geometrically nonlinear finite element analysis are shown in Figures 7.3(b) and (c) respectively. It shows that the topologies using linear and nonlinear analysis could be quite different. Obviously, the optimal topologies for structures undergoing large deformation depend strongly on the magnitude of the maximum design load.

In order to find out which design is better, the complementary work is calculated using nonlinear finite element analysis for both linear and nonlinear designs. The results are given in Table 7.1. It is seen that designs using geometrically nonlinear finite element analysis are always stiffer than that using linear finite element analysis. However, these improvements are marginal for structures without experiencing snap-through, as discussed by Buhl *et al.* (2000). Table 7.1 also compares the results of nonlinear designs using SIMP method (Buhl *et al.* 2000). It is noted that the BESO designs have slightly lower complementary work, although the topologies from BESO and SIMP methods are very similar. The difference in the values of the complementary work can be attributed to the effect of grey areas in the SIMP topologies where the strain energy of the intermediate density elements may have been overestimated.

Figure 7.2   Design domain, loading and support conditions for a geometrically nonlinear structure.

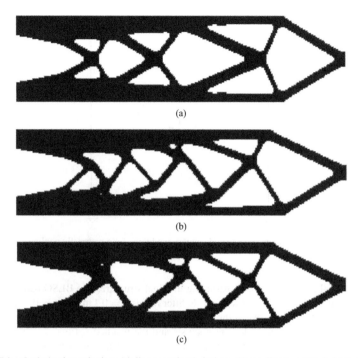

(a)

(b)

(c)

Figure 7.3 Optimized topologies: (a) linear optimal design; (b) nonlinear optimal design for $F = 60$ kN; (c) nonlinear optimal design for $F = 144$ kN.

7.6.2   Materially Nonlinear Structure

In this example, we consider stiffness optimization of the structure shown in Figure 7.4. The dimensions of the plate are 2 m $\times$ 2 m $\times$ 0.01 m. The maximum design load is 20 kN. The structure is made of a frictional material such as soils or rock which exhibits pressure-dependent yield (the material becomes stronger as the pressure increases). Thus a linear Drucker-Prager elastic-perfectly-plastic model with friction angle $\beta = 40°$ and dilation angle $\psi = 40°$ is employed. The solid material has yield stress in uniaxial compression $\sigma_y = 40$ MPa, Young's modulus $E = 20$ GPa and Poisson's ratio $\nu = 0.3$. Assume that only 20 % of design domain

Table 7.1   Comparison of complementary work W^C.

	W^C (kJ)	
Maximum design load	60 kN	144 kN
Linear design from BESO	2.183	12.53
Nonlinear designs from BESO	2.171	12.38
Nonlinear designs in Buhl *et al.* (2000)	2.331	13.29

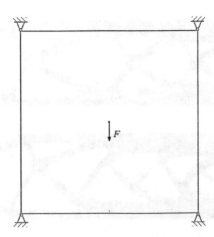

Figure 7.4   Design domain, loading and support conditions for a materially nonlinear structure.

volume material is available for constructing the final structure. The BESO parameters used in this example are $ER = 2\,\%$, $AR_{max} = 2\,\%$, filter radius $r = 0.1$ m and $\tau = 0.1\,\%$.

To show the difference between the linear design and elastoplastic design, the problem is solved first using linear finite element analysis and the resulted optimal design is given in Figure 7.5(a). Since linear material features symmetrical tension and compression behaviour, the linear design also shows symmetric tensile and compressive members. When the problem is solved using nonlinear finite element analysis, a different optimal topology is obtained as shown in Figure 7.5(b). The nonlinear design has effectively taken into account the pressure-dependent strength of the material and thereby produced a structure that is dominated by

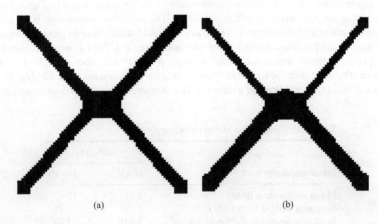

(a)   (b)

Figure 7.5   Optimized topologies: (a) linear optimal design; (b) materially nonlinear optimal design.

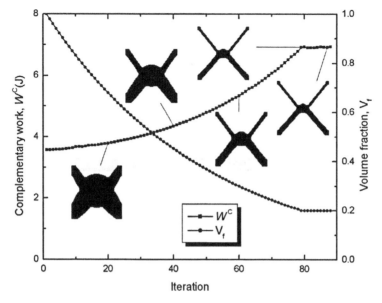

Figure 7.6   Evolution histories of complementary work, topology and volume fraction.

two compressive members. This is certainly an efficient use of the material which becomes stronger when the confining (compressive) stress is higher.

When the nonlinear finite element analysis is applied to both linear and nonlinear designs, it is found that the complementary work and the final deflection are 11.38 J and 0.94 mm for the linear design and 6.93 J and 0.78 mm for the nonlinear design, which means that the nonlinear design is much stiffer than the linear design. Figure 7.6 shows the evolution histories of the complementary work, topology and volume fraction for the nonlinear design. It is seen that the objective function, i.e. the complementary work, has stably converged to a constant value at the final stage. Also, the topology has remained almost the same during the final iterations.

7.6.3   Geometrically and Materially Nonlinear Structures

The present BESO method can be applied directly to both geometrically and materially nonlinear structures. If the optimization process starts from the full design, the cost of nonlinear finite element analysis could be very high, especially for a large 3D structure. To improve the computational efficiency, one may start BESO from an initial guess design with its material volume equal to the objective volume. The optimal design is achieved by redistributing the material (i.e. relocating the solid elements). Since only a small portion of all elements in the design domain is included in nonlinear finite element analysis, the computation time would be significantly reduced.

The structure shown in Figure 7.7(a) is clamped on both sides with a concentrated force, 10 kN, applied to the centre of the top edge. It is 4 m long, 1 m wide and 0.01 m thick. The

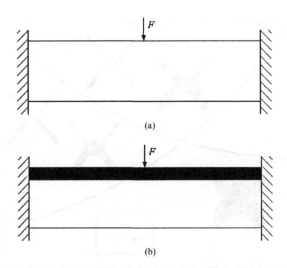

Figure 7.7 A structure with both material and geometrical nonlinearities: (a) design domain; (b) initial guess design.

nonlinear material is assumed to follow the well-known Ramberg-Osgood plasticity model with Young's modulus $E = 500$ MPa, Poisson's ratio $\nu = 0.3$, the yield stress $\sigma_y = 1$ MPa, the yield offset $\alpha = 0.002$ and the hardening exponent $n = 3$. The objective volume is 20 % of design domain. The initial guess design is shown in Figure 7.7(b). The BESO parameters are $ER = 0$, $AR_{max} = 2$ %, $r = 0.05$ m and $\tau = 0.01$ %. The convergence tolerance τ used here is very small to ensure that the BESO algorithm would truly converge to a stationary point.

The optimal design using the linear finite element analysis is shown in Figure 7.8(a). Figure 7.8(b) gives the optimal design using the nonlinear finite element analysis. The two designs are completely different. By applying nonlinear finite element analysis to both linear and nonlinear designs, it is found that the corresponding complementary work and deflection are 633.6 J and 0.11 m for the linear design and 456.9 J and 0.09 m for the nonlinear design respectively. It proves that the nonlinear design is much stiffer than the linear design. Figure 7.9 shows the evolution histories of the complementary work and the structural topology while the material volume is kept constant during the optimization process. After 230 iterations, the solution stably converges to an optimum. However, designs with the complementary work slightly above that of the optimum can be obtained using a larger convergence tolerance τ in far less iterations (e.g. when $\tau = 1$ %, the number of iterations is about 130).

7.6.4   Effects of the Nonlinear Properties and the Magnitude of the Design Load

The design domain of a long beam is shown in Figure 7.10. The beam is 1600 mm long, 200 mm deep and 10 mm thick. The design load, $P = 30$ N, is applied at the centre of the bottom edge. An elastoplastic material model with nonlinear strain-hardening is assumed. The

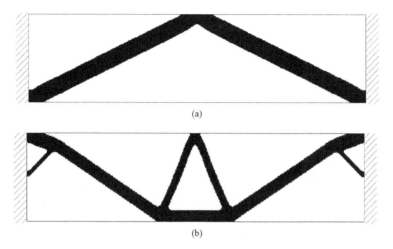

Figure 7.8 Optimized topologies: (a) linear optimal design; (b) geometrically and materially nonlinear optimal design.

stress-strain relationship after yielding is $\sigma = 1.34\varepsilon^{0.5}$. Other properties of the material are Young's modulus $E = 30\,\text{MPa}$, yield stress $\sigma_y = 0.06\,\text{MPa}$ and Poisson's ratio $\nu = 0.3$. The target (objective) material volume is 20 % of the design domain.

In this example, four different optimization cases are considered: linear, geometrically nonlinear only, materially nonlinear only, and both geometrically and materially nonlinear.

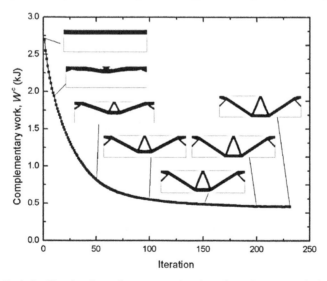

Figure 7.9 Evolution histories of complementary work and topology of the geometrically and materially nonlinear optimal design.

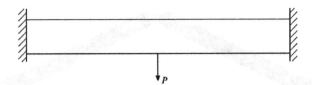

Figure 7.10   Design domain, loading and support conditions of a nonlinear structure.

The design domain is discretized into 640×80 four node quadrilateral elements. The obtained topologies are shown in Figure 7.11 and the iteration numbers are 103, 100, 101 and 99 respectively. It is seen that topologies from the linear and material nonlinear analysis are quite different from the topology from the analysis combining both geometrical and material nonlinearities. However, the topology from geometrically nonlinear analysis is similar to that from the combined nonlinear analysis.

In order to evaluate and compare these different topologies, we re-analyse the four final designs using nonlinear finite element analysis considering both geometrical and material nonlinearities. Figure 7.12 shows the deformed shape of each design under the design load. It is seen that local buckling occurs in the designs obtained from the linear and material nonlinear models. The buckled members significantly reduce the load carrying capability of the structure. Figure 7.13 shows the force-displacement relationships for the four designs. When the applied load is equal to the design load (30 N), the displacement is 112.32 mm for the linear optimum, 118.99 mm for the materially nonlinear optimum, 44.41 mm for geometrically nonlinear optimum and 43.65 mm for the combined nonlinear optimum. It is no surprise that the combined nonlinear optimum is stiffer than other designs. Compared to the linear optimum, the improvement in the stiffness of the combined nonlinear optimum is highly significant.

To investigate the effect of the magnitude of the design load, the above problem is solved for various design loads. The optimization is performed with the combined nonlinear analysis. Four different magnitudes of the design load of 10 N, 15 N, 25 N and 30 N are considered. The obtained optimal topologies are shown in Figure 7.14 and the iteration numbers are 94, 87, 130 and 99 respectively. It is seen that for the small load ($P = 10$ N), the topology for combined nonlinear analysis is similar to the topology obtained from linear analysis. However, as the load increases, the resulting topology changes considerably as the nonlinear effects become more significant.

7.6.5   Three-dimensional Geometrically and Materially Nonlinear Structure

The above BESO method is directly applicable to topology optimization problems of 3D structures. As an example, Figure 7.15(a) shows the maximum design domain, loading and supports conditions of a 3D beam. An elastic, linear hardening plastic model with Young's modulus $E = 1$ GPa, Poisson's ratio $v = 0.3$, yield stress $\sigma_y = 10$ MPa, hardening modulus $E_p = 0.3E$ is assumed for the material. The design domain is meshed using 100 000 hexahedral elements. The objective volume is 5 % of design domain. The initial guess design is shown

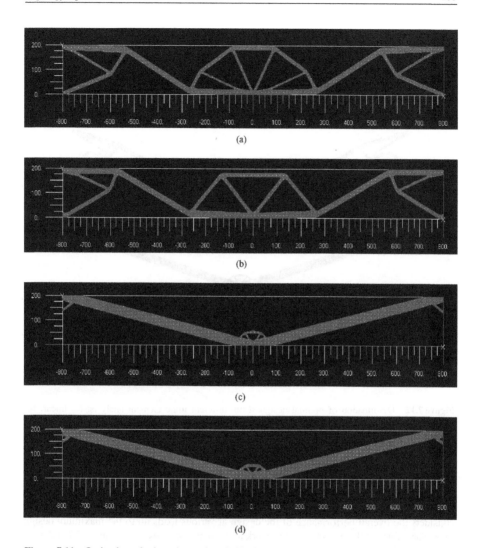

Figure 7.11   Optimal topologies using various finite element analysis: (a) linear; (b) materially nonlinear; (c) geometrically nonlinear; (d) both geometrically and materially nonlinear.

in Figure 7.15(b) with only 5000 solid elements. The BESO parameters are $ER = 0, AR_{max} = 2\%, r = 4\,mm$ and $\tau = 0.01\%$.

Figure 7.16(a) shows the optimal design using linear finite element analysis and Figure 7.16(b) gives the optimal design using geometrically and materially nonlinear finite element analysis. The two topologies are significantly different. By applying geometrically and materially nonlinear finite element analysis to the two topologies shown in Figure 7.16, we obtain

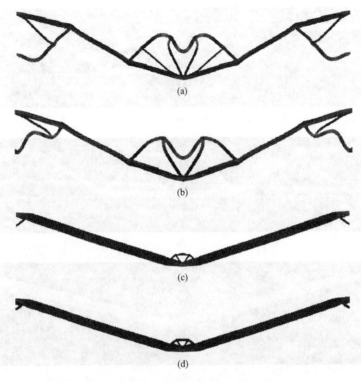

Figure 7.12 Deformation of optimal designs from different finite element analysis: (a) linear; (b) materially nonlinear; (c) geometrically nonlinear; (d) both geometrically and materially nonlinear.

force-displacement relationships for the linear and nonlinear designs, as shown in Figure 7.17. It is found that the topology optimized using nonlinear finite element analysis is stiffer than that using linear finite element analysis at the maximum design load (1.12 kN). However, it has slightly larger deflection for smaller loads. This is because the present optimization procedure considers the overall performance of the design at various loads up to the maximum design load. The values of the complementary work at the maximum design load are 6.7 J and 5.1 J for linear and nonlinear designs respectively. Therefore, we conclude that the nonlinear design is substantially stiffer than the linear design.

7.7   Examples of Nonlinear Structures under Displacement Control

7.7.1   Results from a Relatively Small Design Displacement

A beam 400 mm long and 100 mm high is simply supported at bottom corners as shown in Figure 7.18. A small displacement $d = 5$ mm is applied at the centre of bottom edge. It is assumed that the available material covers 30 % of the design domain. Young's modulus $E = 200$ GPa,

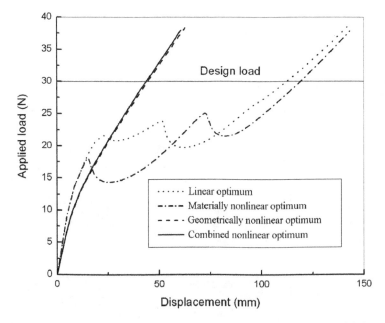

Figure 7.13 Comparison of force-displacement relationships of various optimal designs.

Poisson's ratio $\nu = 0.3$, yield stress $\sigma_y = 300$ MPa and plastic hardening modulus $E_p = 0.3\,E$. To save the computation time, BESO starts from the initial guess design shown in Figure 7.18 with 30 % of the design domain. Thus, only 30 % of elements in the design domain are involved in finite element analysis in each iteration. The BESO parameters are $ER = 0$, $AR_{\max} = 2\,\%$, filter radius $r = 10$ mm, $\tau = 0.01\,\%$ and the maximum iteration number $i_{\max} = 150$.

The topology optimization using linear finite element analysis is first carried out to find the linear design. Figure 7.19 shows the evolution histories of the external work and the topology. It is seen that the external work (nondimensionalized by the external work of the initial guess design) gradually increases until it converges in about 120 iterations. The topology also gradually develops to a stable and mesh-independent design.

When the BESO method is applied to the structure using fully nonlinear finite element analysis, the evolution histories of the external work and the topology are shown in Figure 7.20. It is seen that convergent results for both the external work and the topology are obtained. Due to the fact that the design displacement is relatively small, the final topology of the nonlinear design is similar to that of the linear design although the topologies at early stages of the evolutions are different for the two cases.

In order to evaluate and compare these two designs, the relationships of the applied force and the external work versus the control displacement for both linear and nonlinear designs are determined using fully nonlinear finite element analysis. The results are shown in Figure 7.21. As expected, the nonlinear design has larger external work at the design displacement

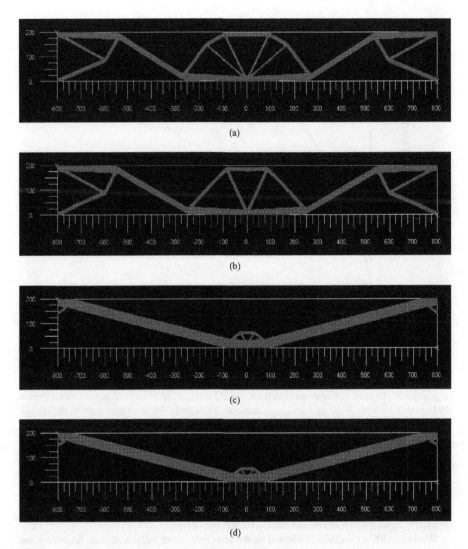

Figure 7.14 Optimal topologies for various design loads: (a) $P = 10$ N; (b) $P = 15$ N; (c) $P = 25$ N; (d) $P = 30$ N.

than the linear design. Furthermore, both the applied force and the external work of the nonlinear design are larger than these of the linear design for most part of the deformation history. However, the nonlinear design is only marginally better than the linear design. This is because the structure is predominantly linear under the small design displacement of 5 mm.

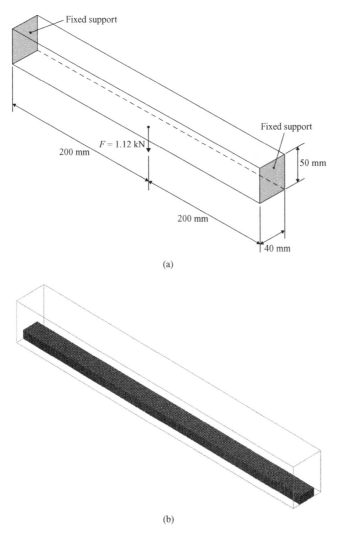

Figure 7.15   Optimization of a 3D structure: (a) design domain; (b) initial guess design.

7.7.2   Results from Large Design Displacements

Here we reconsider the design problem of the previous section with much larger design displacements *d*. Three different design displacements, 20 mm, 50 mm and 100 mm, are applied to the centre of the bottom edge. The resulted evolution histories of the external work are shown in Figure 7.22. In general, the external work in each case tends to increase first and then fluctuates between two bound lines rather than converges to a constant value at the final

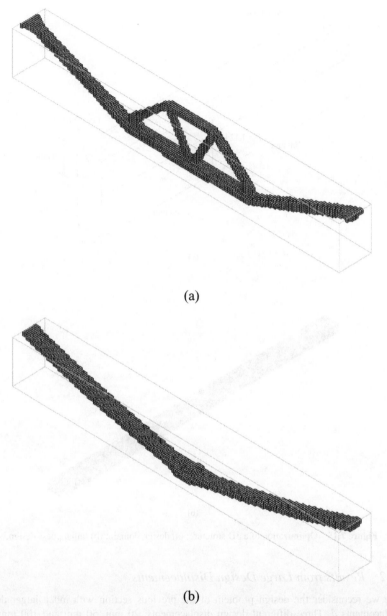

(a)

(b)

Figure 7.16 Optimized topologies: (a) linear design, (b) geometrically and materially nonlinear design.

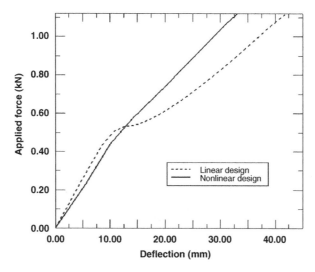

Figure 7.17   Comparison of force-displacement relationships of two optimal designs.

stage. The difficulty in convergence arises mainly because of the alternation of two or more collapse modes of the optimal topology, especially in members experiencing local buckling even though buckling is indirectly taken into account by the nonlinear finite element analysis. Fore this reason, when the design displacement is large it is sometimes impossible for the objective function to converge to a constant value. In such cases, one may select the topology with the highest external work as the optimal design. However, it should be noted that this type of optimal design is sensitive to the variation of the topology and therefore may not be reliable.

Figure 7.23 shows the optimal topologies for design displacements of 20 mm, 50 mm and 100 mm respectively. It is seen that as the design displacement increases the optimal topology changes significantly. The structural members in the topologies for large design displacements have variable cross-sections to effectively resist local buckling. In contrast, the structural members in the linear or small displacement-loaded designs have almost constant cross-sections because there is no local buckling involved. Figure 7.24 shows the deformed

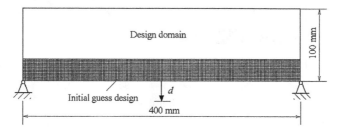

Figure 7.18   Design domain, initial guess design and support conditions for a beam.

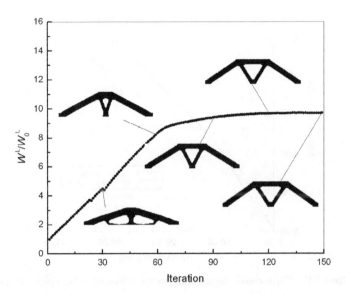

Figure 7.19 Evolution histories of external work and topology for the optimization problem using linear finite element analysis.

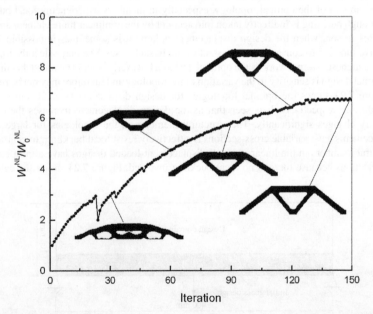

Figure 7.20 Evolution histories of external work and topology for the optimization problem using nonlinear finite element analysis.

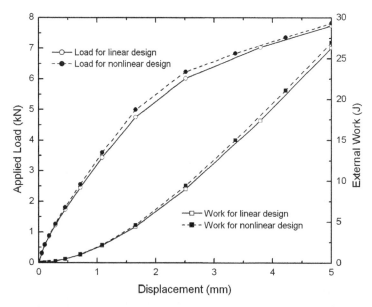

Figure 7.21   Comparison of applied load and external work between linear and nonlinear designs for design displacement $d = 5$ mm.

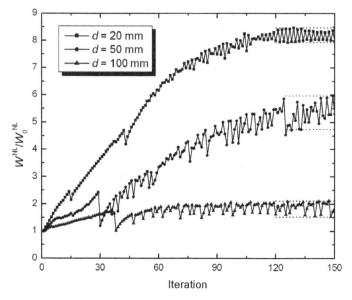

Figure 7.22   Evolution histories of external work for nonlinear structure under large design displacement (the dotted lines indicating the variation bounds of external work at final stage).

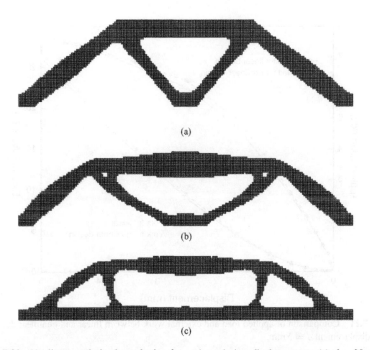

Figure 7.23 Nonlinear optimized topologies for various design displacements: (a) $d = 20$ mm; (b) $d = 50$ mm; (c) $d = 100$ mm.

configurations of the nonlinear designs at their design displacements. It reveals significant differences in the deformation modes of these designs.

Figure 7.25 shows the relationships between the external work and the applied displacement for various designs. Most importantly, it should be noted that the nonlinear design at its design displacement is better than any other designs, e.g. the nonlinear design for $d = 50$ mm has the maximum external work compared to other designs when the displacement reaches 50 mm. The improvement of the nonlinear design over the linear design increases when a larger design displacement is applied. However, the nonlinear design at displacement different from its design displacement can be worse than other designs, e.g. the nonlinear design for $d = 50$ mm has smaller external work than that of the nonlinear design for $d = 20$ mm or the linear design when the displacement reaches 20 mm. It demonstrates that the objective function (external work) also highly depends on the design displacement, and the design displacement should be selected with caution. Figure 7.26 shows the considerable differences in the force-displacement diagrams for various designs. It is interesting to note that the applied force of the design for $d = 100$ mm is much higher than that of the design for $d = 50$ mm when the displacement reaches 100 mm although there is no significant difference between their external works.

7.7.3   Example of a 3D Structure

It should be noted that the most time-consuming part of the optimization process is for solving the equilibrium equations in the finite element analysis. Thus the computational efficiency is

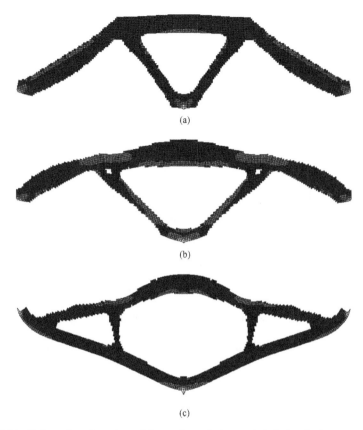

(a)

(b)

(c)

Figure 7.24   Deformed configurations of the nonlinear designs at their design displacements: (a) $d = 20$ mm; (b) $d = 50$ mm; (c) $d = 100$ mm.

of critical importance for large optimization problems, especially for nonlinear 3D structures. To further demonstrate the efficiency of the present BESO method, we consider a 3D structure here. The computation is conducted on a Personal Computer with a 3.0 GHz CPU and 1.0 Gb RAM.

The 3D design domain shown in Figure 7.27 is divided into a $100 \times 10 \times 25$ mesh using 8-node brick elements. A prescribed displacement $d = 50$ mm is applied to three points as shown in Figure 7.27. The nonlinear material is assumed to follow the Ramberg-Osgood plasticity model with Young's modulus $E = 20$ GPa, Poisson's ratio $\nu = 0.4$, the yield stress $\sigma_y = 40$ MPa, the yield offset $\alpha = 0.002$ and the hardening exponent $n = 3$. The objective volume is 20 % of the design domain and the initial guess design with 5000 elements (which satisfies the volume constraint) is also shown in Figure 7.27. The BESO parameters are $ER = 0$, $AR_{\max} = 2$ %, $r = 10$ mm, $\tau = 0.1$ % and $i_{\max} = 100$. The topology is developed by gradually relocating solid elements following the BESO procedure outlined in section 7.5 while the total number of solid elements remains constant.

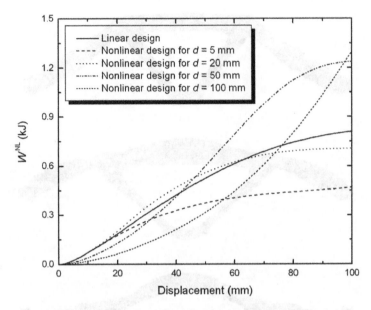

Figure 7.25   Comparison of external work vs displacement for various designs.

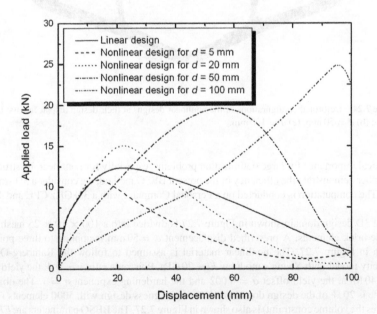

Figure 7.26   Comparison of applied load vs displacement for various designs.

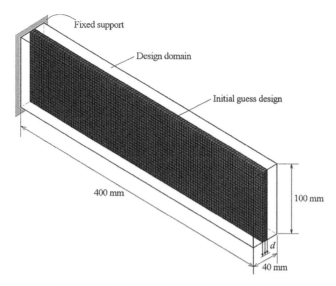

Figure 7.27 Design domain, initial guess design and support conditions for a 3D structure.

The optimal designs from linear and nonlinear finite element analysis are given in Figures 7.28(a) and (b) respectively. The total iteration numbers and the total computation time are 40 and 74.3 minutes for the linear optimization problem, and 29 and 107.3 minutes for the non-linear optimization problem respectively. In this case, obtaining the nonlinear optimal design takes less than two times the computation time spent on the linear design. The main reason for such kind of computational efficiency is that only 20 % of all elements in the design domain are involved in the finite element analysis. We have tested a single finite element analysis on the full design and found that the computation time is 2.5 minutes for the linear analysis, but more than 4 hours for the nonlinear analysis with 40 displacement steps and 162 local iterations. The above analysis clearly demonstrates the computational efficiency of the present BESO method which may start from an initial guess design that is much smaller than the full design.

It is seen from Figure 7.28 that the two optimal topologies are significantly different. To further evaluate these designs, a nonlinear finite analysis is applied to each of the two topologies. The relationships of the applied force (sum of reaction forces at three loading points) and the external work versus the displacement are shown in Figure 7.29. It is seen that the linear design is slightly better than the nonlinear design in the small displacement range, and the nonlinear design is much better than the linear design in the large displacement range. When $d = 50$ mm, the total external work of the nonlinear design is 1.14 kJ which is larger than that of the linear design, 1.06 kJ. The applied force of the nonlinear design is 44.2 kN which is much higher than that of the linear design, 37.5 kN. Under the same displacement loading of $d = 50$ mm, the initial guess design has a total external work of 0.68 kJ and the applied force of 25.4 kN. Therefore, the performance of both linear and nonlinear designs has been substantially improved over the initial guess design although all three designs use exactly the same amount of material.

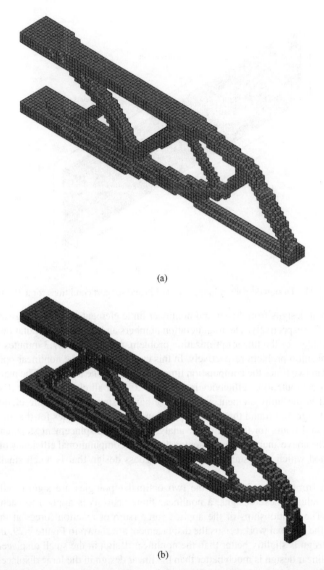

(a)

(b)

Figure 7.28   Optimized topologies from linear and nonlinear finite element analysis: (a) linear design; (b) nonlinear design.

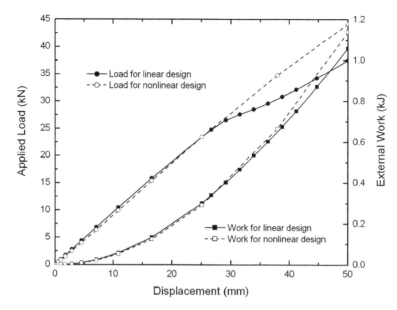

Figure 7.29 Comparison of applied load and external work between the linear and nonlinear designs.

7.8 Conclusion

Topology optimization of nonlinear structures under prescribed load or prescribed displacement using the hard-kill BESO method has been investigated in this chapter. To maximize the stiffness of a structure, the complementary work is minimized for nonlinear structures under the force control, while the external work is maximized for nonlinear structures under the displacement control. In both cases, the sensitivity numbers are equal to elemental strain energies (or strain energy densities for irregular meshes) at the final equilibrium. The optimal design is obtained by gradually removing elements with the lowest sensitivity numbers and adding elements with highest sensitivity numbers.

The developed BESO method has been applied to a number of design problems involving geometrical and/or material nonlinearities. It is found that the optimal designs obtained using nonlinear finite element analysis are stiffer than those using linear finite element analysis, especially when large design loads are applied. Compared to topology optimization of linear structures, even more local optima exist for topology optimization of nonlinear structures. So far, it is almost impossible to guarantee a global optimum for topology optimization problems of nonlinear structures using any numerical methods unless an exhaustive global search is conducted which would be computationally prohibitive. On the other hand, it is required to overcome the convergence difficulties of nonlinear finite element analysis during the evolution of topologies. Considering the complexity of topology optimization problems of nonlinear continuum structures, a practical strategy is to improve the structural performance of a guess design rather than find a global optimum.

We have demonstrated that BESO can start from an initial guess design and find an optimal design by gradually shifting material from one location to another. Since only a small portion of elements in the design domain is included in finite element analysis, the computation time is reduced significantly, especially for large 3D structures. This advantage is extremely important for topology optimization of nonlinear structures.

References

Bendsøe, M.P., Guedes, J.M., Plaxton, S. and Taylor, J.E. (1996). Optimization of structure and material properties for solids composed of softening material. *Int. J. Solids Struct.* **33**(12): 1799–1813.

Bruns, T.E. and Tortorelli, D.A. (2003). An element removal and reintroduction strategy for the topology optimization of structures and compliant mechanisms. *Int. J. Numer. Meth. Engng.* **57**: 1413–30.

Buhl, T., Pedersen, C.B.W. and Sigmund, O. (2000). Stiffness design of geometrically nonlinear structures using topology optimization. *Struct.l Multidisc. Optim.* **19**: 93–104.

Crisfield, M.A. (1991). *Non-linear Finite Element Analysis of Solids and Structures.* New York: John Wiley & Sons, Inc.

Gea, H.C. and Luo, J. (2001). Topology optimization of structures with geometrical nonlinearities. *Comput. Struct.* **79**: 1977–85.

Huang, X. and Xie, Y.M. (2007). Bi-directional evolutionary structural optimization for structures with geometrical and material nonlinearities. *AIAA J.* **45**(1): 308–13.

Huang, X. and Xie, Y.M. (2008). Topology optimization of nonlinear structures under displacement loading. *Engng. Struct.* **30**: 2057–68.

Jung, D. and Gea, H.C. (2004). Topology optimization of nonlinear structures. *Finite Elements in Analysis and Design* **40**: 1417–27.

Pedersen, C.B.W., Buhl, T.E. and Sigmund, O. (2001). Topology synthesis of large-displacement compliant mechanisms. *Int. J. Numer. Meth. Engng.* **50**: 2683–2705.

Pedersen, P. (1998). Some general optimal design results using anisotropic, power law nonlinear elasticity. *Struct. Optim.* **15**: 73–80.

Yuge, K. and Kikuchi, N. (1995). Optimization of a frame structure subjected to a plastic deformation. *Struct. Multidisc. Optim.* **10**: 197–208.

8

Optimal Design of Energy Absorption Structures

8.1 Introduction

Energy absorption structures are employed where collision may cause serious consequences such as injury or fatality to humans and damage to vehicles (Johnson and Reid 1978; 1986; Jones 1989; Lu and Yu 2003). When such a structure is subjected to collision, the external kinetic energy is dissipated, to a great extent, by its large, plastic deformation. Design optimization of energy absorption structures is of special interest in the automotive industry. There has been some research carried out on optimizing parameters (e.g. dimensions) of tubular structures using structural optimization techniques (Lust 1992; Avalle *et al.* 2002; Jansson *et al.* 2003). It is noted, however, that there has been very limited work on *topology* optimization of energy absorption structures despite its great potential. The work by Pedersen (2003; 2004) on crashworthiness design deals with discrete frame structures. In this chapter, we present the more challenging work of topology optimization of continuum structures for energy absorption using a hard-kill BESO method (Huang *et al.* 2007).

8.2 Problem Statement for Optimization of Energy Absorption Structures

Topology optimization problems of energy absorption structures usually have certain constraints, such as limits on the force and the deformation (see Figure 8.1). Typically, a maximum allowable crushing distance is prescribed, so as to retain sufficient space for survival of the occupants or other important devices. At the same time, a high level of force is required in order to dissipate a large amount of energy. However, the maximum force should not be too large, as it might be beyond the tolerance level of the occupants. In other words, the energy absorption structure should be neither too stiff, which may exceed the force limit; nor too compliant, which may exceed the allowable crushing distance. Therefore, an ideal

Evolutionary Topology Optimization of Continuum Structures: Methods and Applications   Xiaodong Huang and Mike Xie
© 2010 John Wiley & Sons, Ltd

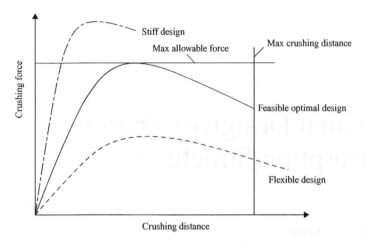

Figure 8.1 Typical load-displacement curve and design constraints for energy absorption structure.

energy absorption structure should possess a rectangular force-displacement relationship (see Figure 8.1), although this is practically unachievable.

To obtain the most efficient energy absorption design, one may maximize the total absorbed energy per unit volume (E/V) within the prescribed limits for the force and the displacement. Thus, the optimization problem can be formulated using the elements as the design variables as

$$\text{Maximize} \quad f(x) = \frac{E}{V} \tag{8.1a}$$

$$\text{Subject to} \quad F_{max} = F^* \tag{8.1b}$$

$$U_{max} = U^* \tag{8.1c}$$

$$x_j \in \{0, 1\} \quad j = 1, \cdots, M \tag{8.1d}$$

where F is the external force and U is the displacement. F^* and U^* are the allowable maximum force and displacement. The binary design variable x_j declares the absence (0) or presence (1) of an element. M is the total number of elements in the design domain.

To simulate the crushing behaviour of a structure, nonlinear finite element analysis is conducted by gradually increasing the displacement of impact points from 0 to the maximum allowable crushing distance, U^*. Therefore, the maximum displacement constraint of Equation (8.1c) is easily satisfied. To satisfy the maximum force constraint, the sensitivity information of the maximum force needs to be calculated in conventional optimization methods. However, this is a difficult task for energy absorption structures with geometrical and material nonlinearities. In this chapter it is assumed that the force-displacement curve does not change significantly before and after an element is eliminated from the design domain as shown in Figure 8.2 and that the maximum force will decrease or increase as the total volume of the structure decreases

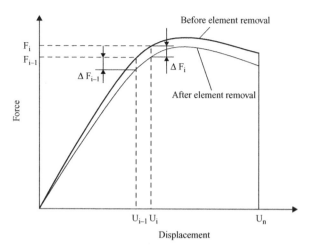

Figure 8.2 Force-displacement curves before and after removing an element for sensitivity analysis.

or increases. Therefore, the force constraint of Equation (8.1b) will be heuristically satisfied by varying the total volume of the structure. As a result, only the sensitivity of the objective function in Equation (8.1a) needs to be considered. This will be discussed below.

8.3 Sensitivity Number

8.3.1 Criterion 1: Sensitivity Number for the End Displacement

The variation of the objective function with respect to the change in design variable x is

$$\Delta f(x) = \frac{1}{V}\left(\Delta E - \frac{E}{V}\Delta V\right) \tag{8.2}$$

According to the principle of energy conservation, the total strain energy is equal to the external work. When the structure is crushed to the end displacement $\mathbf{U}^*$ (which is equal to $\mathbf{U}_n$), the total strain energy E and the total external work W can be obtained from Figure 8.2 as

$$E = W = \lim_{n \to \infty} \frac{1}{2} \sum_{i=1}^{n} \left[\left(\mathbf{U}_i^T - \mathbf{U}_{i-1}^T\right)\left(\mathbf{F}_i + \mathbf{F}_{i-1}\right)\right] \tag{8.3}$$

It is noted that the above expression is identical to the defined mean compliance of nonlinear structures in Chapter 7 (see Equation 7.2).

When the jth element is completely removed from the system, according to the sensitivity analysis in Chapter 7, the variation of the external work can be approximately expressed by the final strain energy of the jth element as

$$\Delta E = -E_n^j \tag{8.4}$$

where E_n^j is the total strain energy of the jth element when $\mathbf{U} = \mathbf{U}^*$. Meanwhile, the variation of the total volume can be easily calculated by

$$\Delta V = -V_j \tag{8.5}$$

where V_j denotes the volume of the jth element. Substituting Equations (8.4) and (8.5) into Equation (8.2), the variation of the objective function can be rewritten as

$$\Delta f(x) = \frac{1}{V}\left(\frac{V_j}{V}E - E_n^j\right) \tag{8.6}$$

From the above equation, a nondimensional elemental sensitivity number is defined by dividing the variation of the objective function by E/V as

$$\alpha_n^j = \frac{V_j}{V} - \frac{E_n^j}{E} \tag{8.7}$$

The elemental sensitivity number provides an estimate of the relative ranking of each element in terms of its effect on the objective function if it is removed. Note that the elemental sensitivity number can be positive or negative, which implies that the objective function may decrease or increase when an element is removed. To maximize the objective function, those solid elements with the highest positive sensitivity numbers should be switched to void; at the same time, those void elements with the lowest negative values should be changed to solid.

8.3.2   Criterion 2: Sensitivity Number for the Whole Displacement History

In general, the actual crushing distance of a structure varies with the amount of external kinetic energy and may not always reach the maximum allowable displacement $\mathbf{U}^*$ in a collision. In order to produce an efficient design that is suitable for any magnitude of impact within the allowable limit, the above definition of the sensitivity number can be extended to multiple crushing distance cases. For different crushing distances, the removal of an element has different effects on the objective function. An overall sensitivity number should be defined, which can be used to estimate the effect of the removal of an element on the external works for all crushing distance cases. Thus

$$\alpha^j = \sum_{i=1}^{n} \alpha_i^j \tag{8.8}$$

where n is the number of crushing distance cases. The above sensitivity number can be found by gradually increasing the crushing distance from 0 to the maximum. In this chapter, we divide the maximum crushing distance into 10 even design points. The sensitivity number for each case is calculated once the crushing distance reaches the corresponding design point. Then, the topology evolves according to Equation (8.8). As a result, the final topology has high efficiency for absorbing energy across the whole range of the design points.

It is seen that criterion 1 focuses on the maximum energy absorbed per unit volume at the end displacement; whereas criterion 2 considers the energy absorption in the whole deformation history. For criterion 1, the aim is to absorb the largest amount of energy (at the end displacement) with the least amount of material. Therefore, the design performance can be measured by

$$e_1 = \frac{W}{V} \tag{8.9}$$

To meet criterion 2 most effectively, the structure should consistently sustain a high level of load below the allowable maximum crushing force throughout the whole displacement history. In this case, the design performance can be measured by

$$e_2 = \frac{W}{W_{\max}} \tag{8.10}$$

where $W_{\max} = F^* U^*$ is the maximum absorbed energy of an ideal energy absorption structure.

8.4 Evolutionary Procedure for Removing and Adding Material

The evolutionary iteration procedure using a hard-kill BESO method is given below.

Step 1: Discretize the design domain using a fine mesh of finite elements and assign initial property values (0 or 1) to elements to construct an initial design.

Step 2: Carry out nonlinear finite element analysis using FEA software, such as ABAQUS.

Step 3: Check whether or not the optimum has been reached. If so, output the results and terminate the iteration. Otherwise, go to Step 4.

Step 4: Calculate the strain energy of each node and then filter nodal strain energy into each element in the whole design domain.

Step 5: Calculate the elemental sensitivity numbers and average the sensitivity numbers with their historical information.

Step 6: Determine the target volume for the next iteration by comparing the maximum reaction force with the maximum allowable crushing force. In order to satisfy the maximum crushing force constraint, the material volume would be decreased when the maximum crushing force is greater than the maximum allowable crushing force as follows:

$$V_{k+1} = V_k(1 - ER) \quad (k = 1, 2, 3 \cdots) \tag{8.11a}$$

where ER is the evolutionary volume ratio. $ER = 1\%$ is used throughout this chapter. Similarly, the volume would be increased when the maximum crushing force is less than the maximum allowable crushing force as follows

$$V_{k+1} = V_k(1 + ER) \quad (k = 1, 2, 3 \cdots) \tag{8.11b}$$

Step 7: Reset the property values of elements. For solid elements, the property value is switched from 1 to 0 if

$$\alpha_i \geq \alpha_{del}^{th} \qquad (8.12a)$$

For void elements, the property value is switched from 0 to 1 if

$$\alpha_i < \alpha_{add}^{th} \qquad (8.12b)$$

where α_{del}^{th} and α_{add}^{th} are the threshold sensitivity numbers for removing and adding elements.

Step 8: Construct a new design with updated solid elements and check the boundary and loading conditions. Go back to Step 2.

8.5 Numerical Examples and Discussions

8.5.1 Example 1

The simply supported structure shown in Figure 8.3 is 100 mm long, 20 mm deep and 1 mm thick. The crushing displacement loading is applied at the centre of the top edge. The allowable maximum crushing force is set to be 20 kN and the maximum crushing distance u_{max} is 20 mm. The material has Young's modulus $E = 200$ GPa, Poisson's ratio $v = 0.3$, yield stress $\sigma_y = 300$ MPa and plastic hardening modulus $E_p = 0.3\ E$. The design domain is discretized using 200×40 four node plane stress elements (which do not consider out-of-plane buckling). The mesh-independency filter radius is 3 mm.

In this example, BESO starts from the initial full design. Figures 8.4(a) and (b) show the evolution histories of the volume fraction, absorbed energy per unit volume and the maximum crushing force using criteria 1 and 2 respectively. In both cases, the absorbed energy per unit volume tends to increase as the volume decreases, and the maximum crushing force gradually decreases as the volume decreases until it satisfies the force constraint. It should be pointed out that the sudden jumps in the absorbed energy per unit volume in Figure 8.4(a) are caused

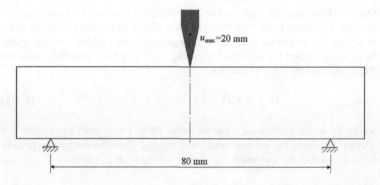

Figure 8.3   Design domain, displacement loading and support conditions for example.

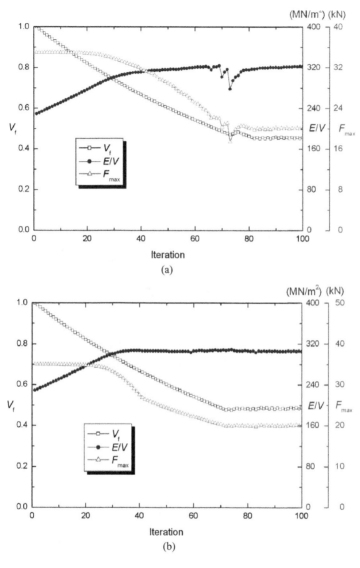

Figure 8.4 Evolution histories of volume fraction V_f, absorbed energy per unit volume E/V and maximum crush force $F_{\max}$ for example 1 using different criteria: (a) criterion 1; (b) criterion 2.

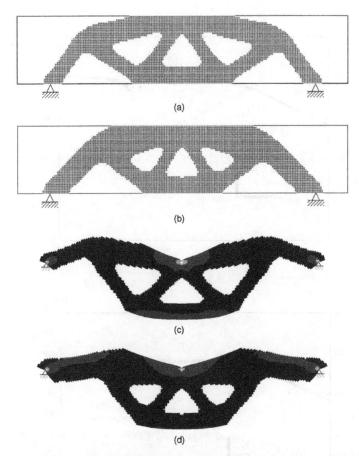

Figure 8.5   Optimal designs and their deformed shapes: (a) optimal design using criterion 1; (b) optimal design using criterion 2; (c) final deformation of optimal design (a); (d) final deformation of optimal design (b).

by significant changes of topology as a result of bar eliminations. Thereafter, the absorbed energy per unit volume recovers and the topology develops in the right direction. At the final stage of the evolution, the absorbed energy per unit volume converges to a maximum value while the volume oscillates within a narrow band of 1 % of the current volume as the force constraint is satisfied and violated alternately in two successive iterations.

The obtained optimal designs and their final deformation are given in Figure 8.5. Figure 8.6(a) shows the force-displacement curves of the optimal designs and the initial full design. It is seen that the initial full design does not satisfy the force constraint as the force exceeds the maximum allowable value of 20 kN long before the prescribed maximum crushing distance of 20 mm is reached. In contrast, the maximum crushing forces of both optimal designs obtained from the BESO method are within the allowable limit. Figure 8.6(b) shows the energy-displacement curves for the two optimal designs. It is seen that the design

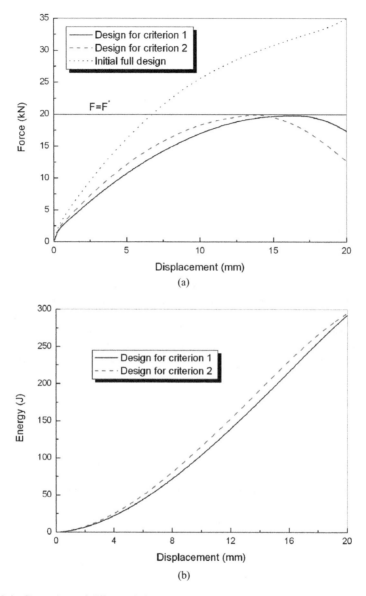

Figure 8.6 Comparison of different designs: (a) force-displacement curves of optimal designs and initial full design; (b) energy-displacement curves of optimal designs.

Table 8.1 Detailed comparison of various designs from four examples.

	Design	Volume fraction, $V_f(\%)$	F_{max}	F^*	W (J)	$e_1(MN/m^2)$	e_2
Example 1	Initial full design	100	35.02 kN	20 kN	$-^1$	229.16	–
	Design for criterion 1	45.7	19.98 kN	20 kN	289.70	316.96	0.724
	Design for criterion 2	48.5	19.92 kN	20 kN	295.44	304.58	0.739
Example 2	Initial full design	100	35.64 kN	20 kN	–	174.78	–
	Design for criterion 1	41.9	19.96 kN	20 kN	222.27	265.24	0.741
	Design for criterion 2	45.1	19.91 kN	20 kN	227.00	251.66	0.757
Example 3	Initial full design	100	102.82 N	70 N	–	180.79	–
	Design for criterion 1	51.8	69.58 N	70 N	0.7045	226.67	0.503
	Design for criterion 2	56.9	69.99 N	70 N	0.7436	217.81	0.531
Example 4	Initial guess design	50	29.25 N	70 N	0.2568	85.59	0.183
	Design for criterion 1	51.8	70.0 N	70 N	0.7018	225.89	0.501
	Design for criterion 2	55.0	69.55 N	70 N	0.7310	221.50	0.522

[1] Value has no meaning as the design is unacceptable because $F_{max} > F^*$.

using criterion 2 absorbs slightly more energy than the design using criterion 1. This is mainly due to the fact that the former design is heavier than the latter, as shown in Table 8.1.

In Table 8.1 we compare the optimal designs and the initial design in various aspects. Firstly, it is noted that the absorbed energy per unit volume e_1 of each of the two optimal designs has significantly increased compared to the initial full design. The design using criterion 1 has the highest e_1 among the three designs. These results clearly demonstrate the effectiveness of the present BESO procedure for improving the energy absorption performance of structures. Secondly, the design using criterion 2 has higher e_2 (as well as higher volume) than the design using criterion 1. Note that e_2 is a measure of the total energy absorbed by the structure.

8.5.2 Example 2

In Example 2, the same design domain, supporting conditions and material as in Example 1 are used, but the structure is crushed at different locations as shown in Figure 8.7. The allowable maximum force is 20 kN and the maximum crushing distance is 15 mm.

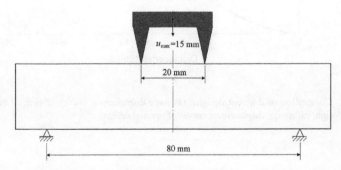

Figure 8.7 Design domain, displacement loading and support conditions for example.

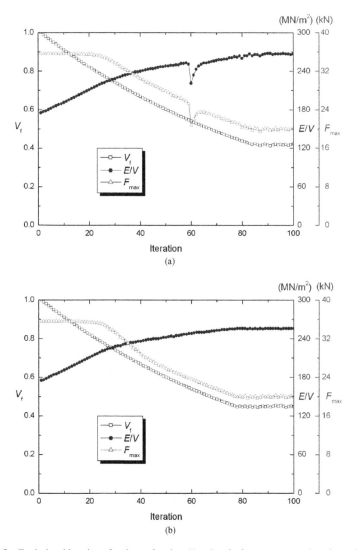

Figure 8.8 Evolution histories of volume fraction V_f, absorbed energy per unit volume E/V and maximum crush force F_{max} for example 2 using different criteria: (a) criterion 1; (b) criterion 2.

The evolution histories of the volume fraction, the absorbed energy per unit volume and the maximum crushing force are shown in Figure 8.8(a) for criterion 1 and in Figure 8.8(b) for criterion 2. These results indicate that optimal designs (at least local optimal designs) for both criteria have been obtained because the absorbed energy per unit volume converges to a maximum value at the final stage. Figure 8.9 shows the obtained topologies and their final deformation. Compared to the topologies in Figure 8.5, the two bars in the middle are

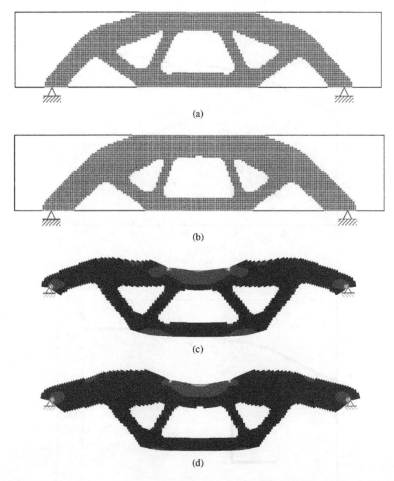

(a)

(b)

(c)

(d)

Figure 8.9 Optimal designs and their deformed shapes: (a) optimal design using criterion 1; (b) optimal design using criterion 2; (c) final deformation of optimal design (a); (d) final deformation of optimal design (b).

now moved apart to support the two separate external loads. The force-displacement curves in Figure 8.10(a) illustrate that both optimal designs satisfy the force constraint whereas the initial full design does not. Figure 8.10(b) shows the energy-displacement curves of both optimal designs. More details about these designs are given in Table 8.1. Once again, the energy absorbed per unit volume of each of the optimal designs has been significantly improved compared to the initial full design. The design using criterion 1 is the best in terms of the energy absorbed per unit volume, e_1. The design using criterion 2 has higher total absorbed energy, or e_2, (as well as higher volume) than the design using criterion 1.

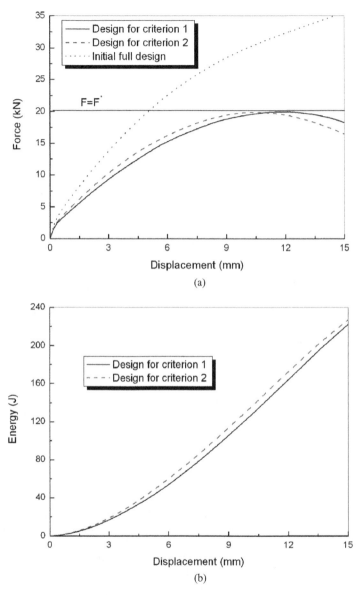

Figure 8.10 Comparison of different designs: (a) force-displacement curves of optimal designs and initial full design; (b) energy-displacement curves of optimal designs.

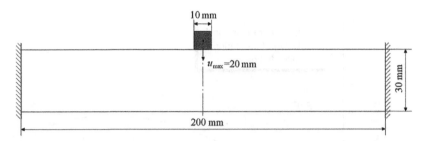

Figure 8.11   Design domain, displacement loading and support conditions for example.

8.5.3   Example 3

In the third example, a structure, 200 mm long, 30 mm high and 1 mm thick, is fixed along both ends. A rigid object collides with the structure at the centre of the top edge as shown in Figure 8.11. The material of the structure has Young's modulus $E = 1$ GPa, Poisson's ratio $\nu = 0.3$, yield stress $\sigma_y = 1$ MPa and plastic hardening modulus $E_p = 0.1\ E$. The allowable maximum crushing force and crushing distance are set to be 70 N and 20 mm respectively. The design domain is discretized using 400×60 four node plane stress elements. The mesh-independency filter radius is 3 mm.

Figure 8.12 shows the evolution histories of the volume fraction, the absorbed energy per unit volume and the maximum crushing force. The optimal designs obtained from using criteria 1 and 2 and their final deformation are presented in Figure 8.13. Figure 8.14(a) shows the force-displacement curves of the optimal designs and the initial full design. It is seen that both optimal designs satisfy the force constraint whereas the initial full design does not. The energy-displacement curves for both optimal designs are shown in Figure 8.14(b). Table 8.1 provides more details about these designs. Similar conclusions can be drawn as in previous examples, i.e. both optimal designs are significantly better than the initial full design in terms of energy absorbed per unit volume and the design using criterion 2 (which has more material) absorbs more energy than the design using criterion 1.

8.5.4   Example 4

To save the computation time, we may begin the BESO process from an initial guess design. In this example we solve the problem in example 3 by starting from the initial guess design shown in Figure 8.15, which occupies 50 % of the full design domain. Figures 8.16(a) and (b) show evolution histories of the volume fraction, the absorbed energy per unit volume and the maximum crushing force using criteria 1 and 2 respectively. It is seen that the absorbed energy per unit volume increases to a maximum value as the topology evolves. Meanwhile, the volume increases and then decreases as the maximum crushing force falls below and rises above the maximum allowable value. Similar to the previous examples, the absorbed energy per unit volume converges to a maximum value at the final stage, and the volume oscillates within a narrow band of 1 % of the current volume as the force constraint is satisfied and violated alternately in two successive iterations.

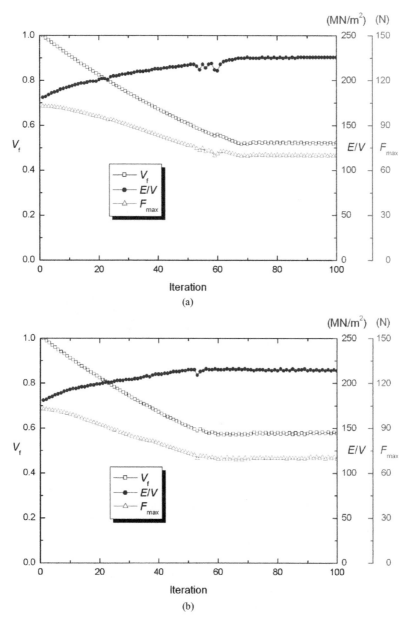

Figure 8.12 Evolution histories of volume fraction V_f, absorbed energy per unit volume E/V and maximum crush force F_{max} for example 3 using different criteria: (a) criterion 1; (b) criterion 2.

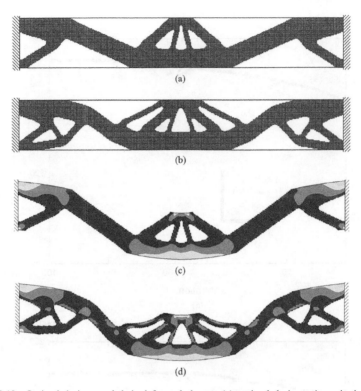

Figure 8.13 Optimal designs and their deformed shapes: (a) optimal design using criterion 1; (b) optimal design using criterion 2; (c) final deformation of optimal design (a); (d) final deformation of optimal design (b).

The optimal designs obtained from using criteria 1 and 2 are given in Figure 8.17 together with their final deformation. Compared to the topology results shown in Figure 8.13 from the initial full design, the optimal design using criterion 1 is almost identical whereas the optimal design using criterion 2 is slightly different. Figure 8.18(a) shows the force-displacement curves of the optimal designs and the initial guess design. It is seen that the two optimal designs satisfy the force constraint ($F_{max} = F^*$) whereas the maximum crushing force of the initial guess design is far below the allowable limit. As shown in Figure 8.18(b) and Table 8.1, although the three designs use similar amount of material, each of the two optimal designs can absorb almost three times as much energy as the initial guess design. This highlights the substantial benefit that can be achieved from applying topology optimization techniques to the design of energy absorption structures.

A more detailed comparison of these designs is given in Table 8.1. The results are similar to those of Example 3. However, the computation time for example 4 is significantly less because the costly nonlinear finite element analysis is conducted on much smaller models.

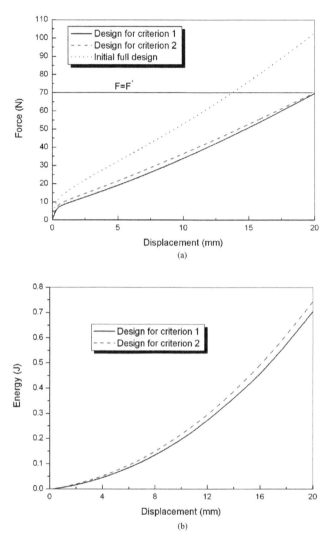

Figure 8.14   Comparison of different designs: (a) force-displacement curves of optimal designs and initial full design; (b) energy-displacement curves of optimal designs.

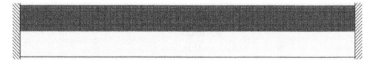

Figure 8.15   Initial guess design for example 4.

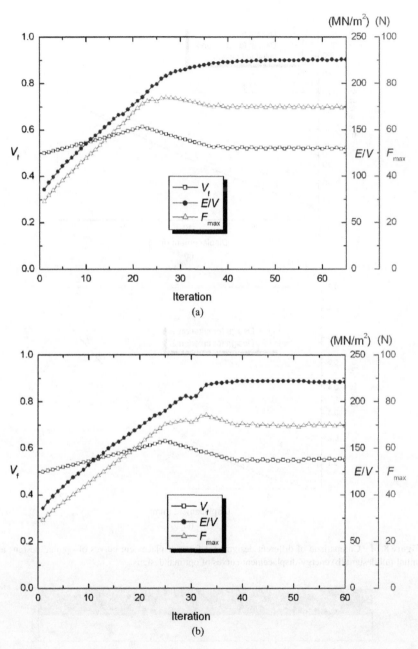

Figure 8.16 Evolution histories of volume fraction V_f, absorbed energy per unit volume E/V and maximum crush force F_{max} for example 4 using different criteria: (a) criterion 1; (b) criterion 2.

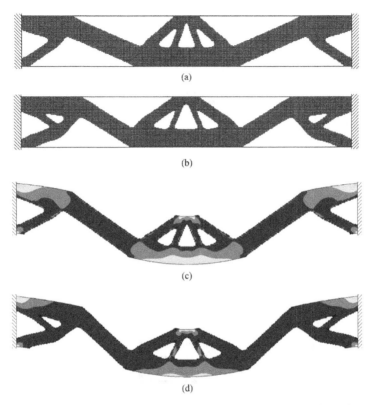

Figure 8.17 Optimal designs and their deformed shapes: (a) optimal design using criterion 1; (b) optimal design using criterion 2; (c) final deformation of optimal design (a); (d) final deformation of optimal design (b).

8.6   Conclusion

In this chapter, a topology optimization procedure for energy absorption structures has been developed using the hard-kill BESO method. Sensitivity numbers for two different criteria are derived in order to maximize the energy absorbed per unit volume at the end displacement or during the whole displacement history. Numerical examples show that an initial full design or a guess design which may not satisfy the force constraint can evolve to an optimal design which satisfies both force and displacement constraints. Compared to the initial full design or the guess design, the optimal design using criterion 1 or 2 leads to significantly improved energy absorption performance measure e_1 or e_2. The results also show that the design using criterion 1 with less material has a higher e_1 than that using criterion 2; whereas, the design using criterion 2 with more material has a higher e_2 than that using criterion 1. The BESO method presented in this chapter is applicable to the topology design of both 2D and 3D continuum structures for the purpose of substantially enhancing or optimizing their energy absorption capabilities.

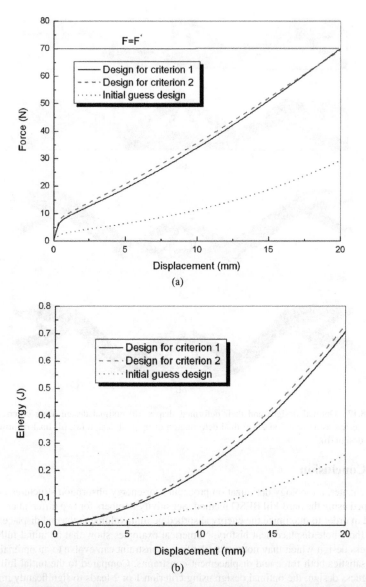

Figure 8.18 Comparison of different designs: (a) force-displacement curves of optimal designs and initial guess design; (b) energy-displacement curves of optimal designs and initial guess design.

References

Avalle, M., Chiandussi, G. and Belingardi, G. (2002). Design optimization by response surface methodology: application to crashworthiness design of vehicle structures. *Struct. Multidisc.Optim.* **24**: 325–32.

Huang, X., Xie, Y.M. and Lu, G. (2007). Topology optimization of energy absorbing structures. *Inter. J. Crashworthiness* **12**(6): 663–75.

Jansson, T., Nilsson, L. and Redhe, M. (2003). Using surrogate models and response surface in structural optimization – with application to crashworthiness design and sheet metal forming. *Struct. Multidisc. Optim.* **25**: 129–40.

Johnson, W. and Reid, S.R. (1978). Metallic energy dissipating systems. *Applied Mechanics Review* **31**: 277–88.

Johnson, W. and Reid, S.R. (1986). Update to 'metallic energy dissipating systems'. *Applied Mechanics Review Update*: 315–19.

Jones, N. (1989). *Structural Impact*. Cambridge University Press.

Lu, G. and Yu, T.X. (2003). *Energy Absorption of Structures and Materials*. Cambridge: Woodhead Publishing.

Lust, R. (1992). Structural optimization with crashworthiness constraints. *Struct. Optim.* **4**: 85–9.

Pedersen, C.B.W. (2003). Topology optimization for crashworthiness of frame structures. *Inter. J. Crashworthiness* **8**(1): 29–39.

Pedersen, C.B.W. (2004). Crashworthiness design of transient frame structures using topology optimization. *Comput. Meth. Appl. Mech. Engng.* **193**: 653–78.

9

Practical Applications

9.1 Introduction

As ESO/BESO methods have reached a level of maturity, more research effort should be directed towards improving its applicability to practical design problems and making the technology easily accessible to practising engineers, architects, and others. One important consideration of any topology optimization technique is the manufacturability of the obtained solution. For example, if we look at the optimal designs given in Tables 5.1 and 5.2, it is quite obvious that the solutions shown in Table 5.1 are practically useless because the details of these designs are too difficult to interpret and too costly to fabricate. In contrast, the results shown in Table 5.2 provide much clearer definitions of the optimal topologies.

As in many other numerical methods of topology optimization, ESO/BESO uses finite elements to model a structure. As a consequence, the topology result from ESO/BESO has jagged edges in the final finite element model of the optimal design. For example, a loaded knee structure is given in Figure 9.1(a). Using the BESO method for compliance minimization, the optimal design shown in Figure 9.1(b) is obtained. For the purpose of practical application, the jagged edges in the finite element model can be easily smoothed out using various image processing techniques such as spline fitting of boundary curves and surfaces. The result is illustrated in Figure 9.1(c). The smoothing technique is computationally inexpensive and simple to implement. It can be readily linked to the ESO/BESO computer code as part of the post-processor of the results.

With the advent of rapid prototyping techniques (e.g. Dimitrov *et al.* 2006), it is now possible to fabricate complicated topologies directly from computer models using 3D printers to reproduce structural details point-by-point and layer-by-layer. Indeed, a technique similar to 3D printing, known as Contour Crafting, is being developed for the automated construction of whole buildings (Khoshnevis 2004). This type of fabrication technology could be used in the future to construct unconventional architectural designs involving complex geometries generated by topology optimization techniques.

In this chapter we present four case studies on practical applications of ESO/BESO methods. The examples discussed range from preliminary conceptual designs to constructed buildings. The first two projects were conducted by a group of Japanese architects and engineers. The remaining examples were from researchers of the Innovative Structures Group at RMIT

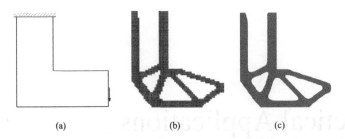

Figure 9.1 Topology optimization of a loaded knee structure: (a) design domain; (b) finite element model of optimal design; (c) smoothed design.

University (www.isg.rmit.edu.au) in collaboration with the Spatial Information Architecture Laboratory (www.sial.rmit.edu.au), Felicetti Pty. Ltd. (www.felicetti.com.au) and BKK Architects (www.b-k-k.com.au).

9.2 Akutagwa River Side Project in Japan

The office building shown in Figure 9.2 was designed using an extended ESO method (Ohmori *et al.* 2005). In essence, the extended ESO method is a BESO method based on the stress level. The building had been planned as part of a large scale redevelopment of a shopping area near Takatsuki JR station in Japan. The land size for the building was approximately 10 m × 6 m.

Figure 9.2 South-west view of an office building designed using BESO (Ohmori *et al.* 2005). Reproduced by permission of Information Processing Society of Japan.

The BESO procedure was applied to the south, west and north side walls simultaneously, while the east side wall and the floor slabs were kept unchanged. In the finite element model, both dead weight in the vertical direction and earthquake loading in the horizontal direction were included. The topology of the three walls evolved as material was gradually removed from regions with low stress and added to areas with high stress. Details of the optimization and design processes were presented by Ohmori *et al.* (2005).

Figure 9.3 shows the inside and outside views of the building soon after the construction was completed in April 2004.

9.3   Florence New Station Project in Italy

By using the same BESO method as that of the above project, Cui *et al.* (2003), in collaboration with the renowned Japanese architect Arata Isozaki, created an extraordinary structure shown in Figure 9.4. This was a proposal for the Florence New Station, having 400 m length, 40 m width and 20 m height where complex facilities were planned in the huge space underneath the upper deck of uniform thickness. During the optimization process, the structure evolved from an initial design of a deck with legs simply supported at the bottom to the final organic form, as shown in Figure 9.5.

The evolving images in Figure 9.5 provide an insightful glimpse into how BESO worked in this example. It is interesting to note that the structural concept of the initial design was completely different from that of the optimal solution. Through the simple BESO process and the exceptional skills of the architects and engineers involved in this project, an ugly duckling of the uninspiring initial design miraculously grew into a beautiful swan in the final design proposal.

Commenting on the structure shown in Figure 9.4, Sasaki (2005) stated that 'The structural elements were optimally formed within a three-dimensional space while satisfying the given design conditions, and the structural shape thus obtained manifests maximum mechanical efficacy with a minimum use of materials.'

Translating such a daring design into a constructible structure required substantial skills and ingenuities. Sasaki (2005) presented detailed designs for the tree trunk-like legs and the top deck using steel pipes, prestressing rods, steel plates, wire mesh, and concrete. He also suggested a construction sequence for the whole structure.

Although the above proposal was not chosen as the winning design for the Florence New Station project, Isozaki, Sasaki and their coworkers used the same approach and similar concept in several other project proposals and eventually won the design for the 250 m long entrance to the Qatar National Convention Centre. The spectacular structure can be seen at www.qatarconvention.com. A team of engineers from Buro Happold have been involved in the challenging task of transforming the organic form into a built artifact. Much of the construction work of the structure, which began in 2004, is now completed.

9.4   Sagrada Família Church in Spain

This work was carried out in 2004 by researchers in the Innovative Structures Group and the Spatial Information Architecture Laboratory at RMIT University. The objective of the project was to explore opportunities for closer collaboration between architects and engineers. One

(a)

(b)

(c)

(d)

(e)

(f)

Figure 9.3 Photos of the office building after completion (Ohmori *et al.* 2005): (a) second floor inside view; (b) another inside view of second floor; (c) first floor inside view; (d) ground floor outside view; (e) west side view; (f) south-west view. Reproduced by permission of Information Processing Society of Japan.

Figure 9.4 Perspective view of proposed Florence New Station (Cui *et al.* 2003; Architect: A. Isozaki). Reproduced by permission of International Association for Shell and Spatial Structures.

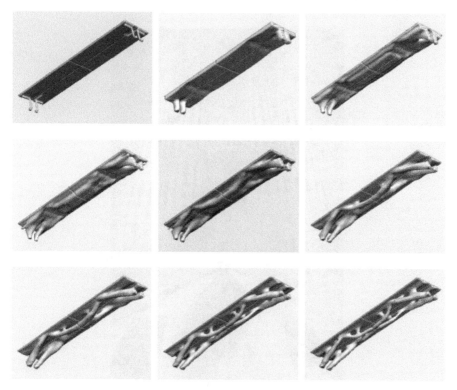

Figure 9.5 Evolution of Florence New Station structure (Cui *et al.* 2003). Reproduced by permission of International Association for Shell and Spatial Structures.

of the case studies conducted by the team was on Antoni Gaudí's Sagrada Familia church in Barcelona, Spain (Burry *et al.* 2005; Felicetti 2009). Gaudí's reference to natural growth and morphogenesis and his use of analogue modelling techniques had much in common with the basic concepts of ESO/BESO. Gaudí famously used cords with sacks of pellets hanging from them to create funicular structural systems and inverted these models upside down to obtain compression-only designs suitable for masonry buildings.

In this study, the ESO method was used to untangle some of the mysteries of Gaudí's design rationale. To find the optimal design for a masonry structure, an ESO algorithm based on the principal stresses was devised in which material with the highest level of tensile stress would be removed iteratively, resulting in a design that was predominantly in compression throughout the structure. Note that it is not always possible to obtain a design that is purely in compression.

The Passion Façade of the Sagrada Familia church was partially completed as shown in Figure 9.6. From a surviving photograph of Gaudí's original drawing for the Passion Façade (Figure 9.7), a sketch shown Figure 9.8 was produced as the starting point for the structural optimization process. The corresponding finite element model of the initial design shown in Figure 9.9 was then generated. Gravity loading was applied to the finite element analysis.

Figure 9.6   Construction of Passion Façade of Sagrada Familia church (Burry *et al.* 2005).

Figure 9.7 Part of surviving photograph of Gaudí's original drawing for Passion Façade (Burry *et al.* 2005).

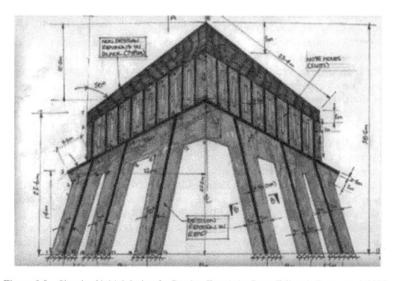

Figure 9.8 Sketch of initial design for Passion Façade by Peter Felicetti (Burry *et al.* 2005).

Figure 9.9   Initial finite element model based on the sketch drawing above (Burry *et al.* 2005).

Figure 9.10 shows the evolution of the façade. The final result could be considered as a structure that would transfer the gravity loading mostly efficiently under the specified conditions for the boundary supports, the amount of material, and the type of material (in this case, masonry). It is noted that each of the lower columns supports a well defined group of branching upper columns, as in Gaudí's original drawing. Interestingly, both the lower and upper columns exhibit bone-like characteristics. Perhaps Gaudí had an unusual insight into optimal structural forms, which is hardly surprising given that he created so many structures which later proved to be highly efficient. It is also possible that he simply drew inspiration from natural load-carrying structures such as bones and shells etc. Apart from structural considerations, the bone-like columns were suggestive of both the tomb and Christ's physical suffering in the crucifixion scene depicted in the sculpture in the porch (see Figure 9.6).

Gaudi's ingenious work on structural optimization was conducted through analogous funicular models under static gravity loading. The digital tools such as ESO/BESO make it much easier to explore a large number of design requirements and options for complex architectural form finding problems. For example, Burry *et al.* (2005) investigated the effect of earthquake loading on the columns for the Passion Façade.

Burry *et al.* (2005) also carried out a series of other studies which revealed remarkable similarities between Gaudi's designs and ESO solutions. Figure 9.11 shows three prototype column models mounted on the upper level of the Passion Façade. These models had been developed from study of Gaudi's original drawing and use of intersecting ruled surfaces by the consultant architect to the Sagrada Familia church, Professor Mark Burry, without any input or influence from the ESO researchers. After these prototype models were created, structural optimization of columns on a sloping surface was performed using a two-dimensional finite

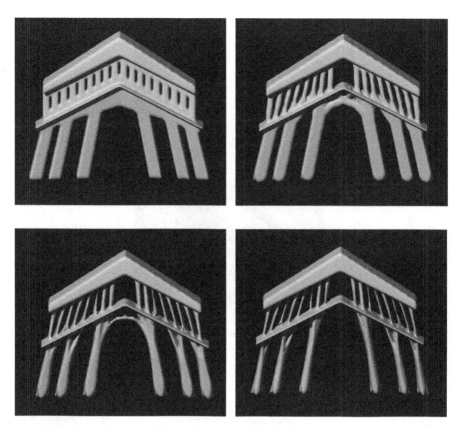

Figure 9.10   Evolution of Passion Façade of Sagrada Familia church (Burry *et al.* 2005).

element model shown in Figure 9.12. The resemblance between the ESO results and the actual columns to be built was amazing. Note that in this case a uniformly distributed vertical load was applied to the top surface, simulating the weight from the gable lintel over the upper level colonnade. Compared to the heavy weight of the large gable lintel, the gravity loading of the columns themselves was negligible. In order to preserve the loading and support conditions, a thin layer of nondesign domain was specified at the top surface and another at the bottom surface.

Burry *et al.* (2005) then applied the same ESO procedure to the three-dimensional model shown in Figure 9.13. Again, the objective was to find a structure that would be in compression. In this example, the top and bottom surfaces were both horizontal. While the bottom surface was fixed to the ground, the top surface was subjected to a uniformly distributed vertical load. In the initial design, there were two narrow necks connecting the top and bottom blocks. After a number of iterations the columns branched out at the top like trees, much akin to the schema of Gaudí's design for the central nave columns shown in Figure 9.14.

Figure 9.11   Three full size prototype column models positioned on Passion Façade (Burry *et al.* 2005).

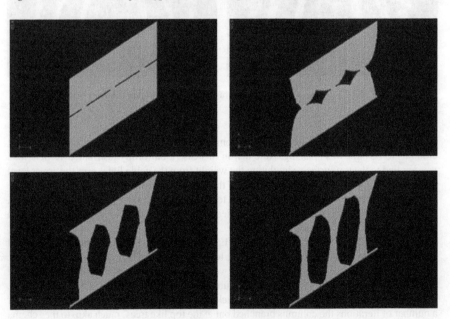

Figure 9.12   Evolution of columns on a sloping surface (Burry *et al.* 2005).

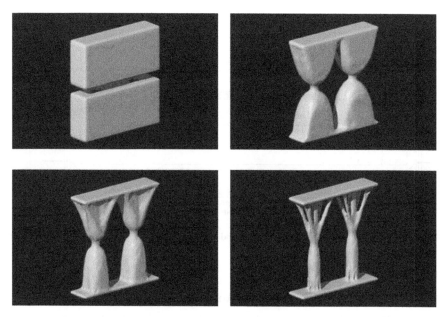

Figure 9.13 Evolution of columns on a horizontal surface (Burry *et al.* 2005).

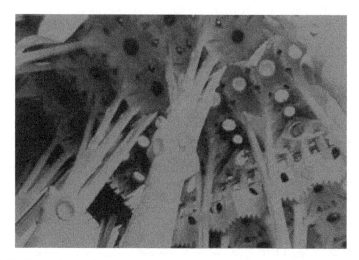

Figure 9.14 Nave columns of Sagrada Familia church with branching elements at the top (Burry *et al.* 2005).

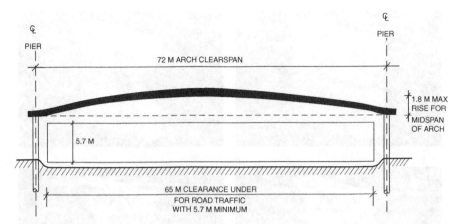

Figure 9.15 Initial sketch from the architect indicating geometrical constraints of the footbridge. Reproduced by permission of BKK Architects.

9.5 Pedestrian Bridge Project in Australia

Recently BKK Architects were commissioned to design a series of pedestrian bridges for a major metropolitan freeway in Australia. The brief for these footbridges called for simple sculptural gestures providing visual interest for the freeway and surrounding environment. Working in collaboration with the Innovative Structures Group, the design team employed the BESO method to create structurally efficient yet strikingly elegant forms.

To start the BESO process, the architect provided a simple sketch shown in Figure 9.15, specifying the geometrical constraints for the bridge. The structure must have a minimum clearance of 65 m width and 5.7 m height, and a maximum ramp slope of 1:20. For the purpose of form finding, only the static loading of 4kPa pressure was applied to the deck of the bridge. Other loading conditions would be considered during the detailed design stage.

The design team explored two different approaches to the BESO process. Firstly they used 3D 'brick' elements in the finite element model. This resulted in the two designs shown in Figure 9.16. Depending on the support conditions at the two ends of the bridge, the optimal topologies were significantly different. Figure 9.16(a) was the solution when no horizontal movement was allowed in the two end planes (i.e. each side of the bridge was supported by a solid vertical abutment). However, if the bridge was supported only at the bottom of the two piers (roller on one side and pin on the other), a truss-like form would emerge, as shown in Figure 9.16(b).

Next the BESO technique for generating periodic structures (Huang and Xie 2008) was utilized to explore a new type of lightweight footbridges in the form of perforated tubes. Figure 9.17 shows the BESO results of the bridge with various cross-sectional shapes (Zuo 2009).

All five designs shown in Figures 9.16 and 9.17 (and a variety of others not presented here) were possible solutions that could be submitted to the client for consideration. Currently BKK Architects and Felicetti are working on a detailed engineering design and the corresponding

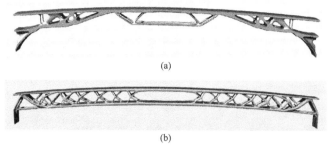

(a)

(b)

Figure 9.16 BESO results from using 3D elements: (a) with no horizontal movement in end planes; (b) with a roller at the bottom of one pier and a pin at the other.

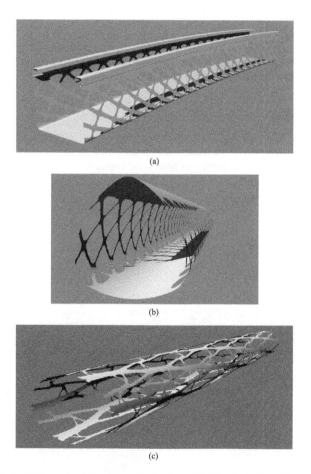

(a)

(b)

(c)

Figure 9.17 Periodically perforated shell topologies for half of the footbridge: (a) rectangular cross-section; (b) egg shaped cross-section; (c) circular cross-section with a twist (Zuo 2009). Reproduced by permission of Mr Zhihao Zuo.

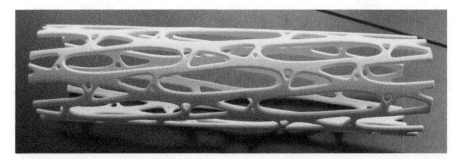

Figure 9.18 Photo of 3D printout of a section of the perforated tube shown in Figure 9.17(c). Reproduced by permission of BKK Architects.

construction techniques for realizing the novel structural form depicted in Figures 9.18 and 9.19 which was derived from the BESO solution shown in Figure 9.17(c).

9.6 Conclusion

The four case studies presented in this chapter on applications of ESO/BESO to practical design problems clearly demonstrate the benefit and potential of using the topology optimization technique as a design tool. It enables architects and engineers to greatly expand the possible

Figure 9.19 Proposal for the footbridge. Reproduced by permission of BKK Architects.

structural forms of their projects and to obtain designs that are not only structurally efficient but also exhibit distinctive aesthetical appeal. The authors believe that one could well see the day when more landmark structures are designed using ESO/BESO and constructed in many parts of the world.

References

Burry, J., Felicetti, P., Tang, J.W., Burry, M.C. and Xie, Y.M. (2005). Dynamical structural modelling – a collaborative design exploration. *Inter. J. Arch. Comput.* **3**(1): 27–42.

Cui, C., Ohmori, H. and Sasaki, M. (2003). Computational morphogenesis of 3D structures by extended ESO method. *J. Inter. Assoc. Shell Spatial Struct.* **44**(1): 51–61.

Dimitrov, D., Schreve, K. and de Beer, N. (2006). Advances in three dimensional printing – state of the art and future perspectives. *Rapid Prototyping J.* **12**(3): 136–47.

Felicetti, P. (2009). Topological revolution. *Architectural Review Australia.* **109**: 50–2.

Huang, X. and Xie, Y.M. (2008). Optimal design of periodic structures using evolutionary topology optimization. *Struct. Multidisc. Optim.* **36**(6): 597–609.

Khoshnevis, B. (2004). Automated construction by contour crafting – related robotics and information technology. *J. of Automation in Construction.* **13**(1): 5–19.

Ohmori, H., Futai, H., Iijima, T., Muto, A. and Hasegawa, H. (2005). Application of computational morphogenesis to structural design. In: *Proceedings of Frontiers of Computational Sciences Symposium*, Nagoya, Japan, 11–13 October, 2005, pp. 45–52.

Sasaki, M. (2005). *Flux Structure*, Tokyo: Toto.

Zuo Z.H. (2009). *Topology Optimization of Periodic Structures*. PhD thesis, RMIT University, Australia.

10

Computer Program BESO2D

Z.H. Zuo
School of Civil, Environmental and Chemical Engineering, RMIT University, Australia

10.1 Introduction

One of the most important advantages of the BESO method is that it can be implemented easily as a post-processor to commercial FEA software packages such as ABAQUS, NASTRAN and ANSYS. However, the reliance on complex and expensive third-party programs severely limits the potential usage and wider applications of the BESO technique. Furthermore, a BESO computer program with its own built-in FEA engine will substantially reduce the data exchange with an otherwise external FEA program and significantly increase the computational efficiency. For these reasons, we have been developing a standalone BESO program for both 2D and 3D structures. Our aim is to make the program as easy to use as possible for engineers, architects and students who may not be familiar with either FEA or structural optimization. This chapter introduces the 2D version of the BESO software called BESO2D. It is provided free of charge for educational purposes. To obtain the program, simply go to the website of the Innovative Structures Group (ISG) at http://isg.rmit.edu.au and download a copy. Updates of BESO2D and other BESO programs can be found from the website or by contacting us via email: Huang.Xiaodong@rmit.edu.au or Mike.Xie@rmit.edu.au.

The current version of BESO2D performs the conventional stiffness optimization (i.e. minimizing the mean compliance, which is the same as maximizing the overall stiffness) of statically loaded structures under plane stress conditions. The program is equipped with functions such as drawing a structure, applying loading and boundary conditions, generating an FEA mesh, performing finite element analysis, and conducting BESO optimization.

10.2 System Requirements and Program Installation

10.2.1 System Requirements

BESO2D is a Windows 32-bit application. Windows XP operating system is recommended, although the program can also be run under other Windows systems such as Windows Vista,

Evolutionary Topology Optimization of Continuum Structures: Methods and Applications   Xiaodong Huang and Mike Xie
© 2010 John Wiley & Sons, Ltd

Windows NT and Windows 98. It is recommended that the computer hardware should have no less than 64 MB RAM and no less than 100 MB hard disk free space for running BESO2D. The actual requirements of the memory and disk space depend on the number of nodes in the FEA model. For a proper display on the screen, a resolution of 1024×768 and a colour setting of 256 colours or above are recommended. Otherwise, the stress distribution image may appear to be of low quality.

Note that BESO2D for other operating systems may be issued in the future and the interested readers can check the ISG website for updates.

10.2.2 Installation of BESO2D

To install BESO2D, follow these steps:

1. Start a computer with a Windows operating system.
2. Connect the computer to the internet and go to the ISG website at http://isg.rmit.edu.au and find the download link of BESO2D. Download the zip file BESO2D.zip to your local hard disk.
3. Create a folder in your hard disk and name it 'BESO2D\'.
4. Unzip BESO2D.zip to the folder 'BESO2D\'. You will find two executable files and some dynamic link libraries (.dll) in the folder.
5. Create a shortcut to BESO2D.exe and copy the shortcut to the Windows Start menu and/or the desktop.
6. Now the installation of BESO2D is complete.

BESO2D can be started by double-clicking BESO2D.exe in the folder 'BESO2D\' or the shortcut in the Windows Start menu (or the desktop). The BESO2D.exe (and its shortcuts) is identified by the icon shown in Figure 10.1.

10.2.3 Constitutive Parts of BESO2D Package

The BESO2D package consists of two executable files: BESO2D.exe and BESO.exe. BESO2D.exe is the main program of the graphic user interface (GUI) window which enables the user to create and view structures graphically. BESO.exe is the engine that performs FEA and structural optimization which is called internally by the GUI window BESO2D.exe. BESO.exe is a self-contained Windows 32-bit console application with minimum interfacing with Windows and can be executed independent of BESO2D.exe in a command line to enhance computational efficiency and to optimize structural models initially created from third-party programs.

Figure 10.1   BESO2D icon.

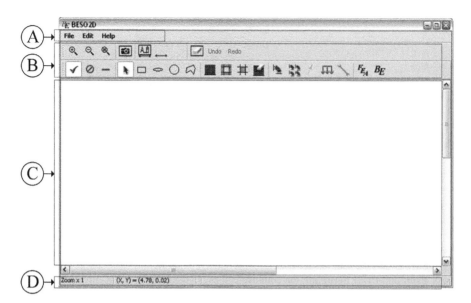

Figure 10.2   The GUI window of BESO2D.

In additional to the two executable files, several dynamic link libraries (dll's) are included which will be called by BESO2D.exe.

10.3   Windows Interface of BESO2D

10.3.1   Overview of the GUI Window

The BESO2D GUI window is shown in Figure 10.2. From top to bottom, the whole interface consists of four parts: (A) the menu bar, (B) the toolbar area, (C) the display area and (D) the status bar.

10.3.2   Menu Bar

Three menus are found in the menu bar (Part A in Figure 10.2): **File**, **Edit** and **Help**. They are explained in the following.

- **File** menu
  There are eight commands in the **File** menu shown in Figure 10.3:
  1. **New** allows the user to start a new design by resetting the current program configurations to their default values and clearing the undo-redo list. The program configurations include the screen width, preselected constraints/loads and material property etc.

Figure 10.3   The File menu.

2. **Open** enables the user to open previously saved program configurations from an existing '.xml' file.
3. **Save** allows the user to save the current program configurations to an '.xml' file.
4. **Print** and **Print Setup** are the commands that are not present in the current version of BESO2D and will be completed in future versions.
5. **Export Model** enables the user to export the current model in the display area to a '.txt' model file.
6. **Import Model** enables the user to import a model from an existing model file.
7. **Exit** is a command to exit the program.
- **Edit** menu
  The **Edit** menu shown in Figure 10.4 contains five commands.
  1. **Cut** and **Copy** are two commands to cut or copy selected constraints or loads (or use the shortcuts 'Ctrl+X'/'Ctrl+C' to perform a cut/copy action).
  2. **Paste** is a command to paste the previously cut or copied items to the display area (or use the shortcut 'Ctrl+V' to perform this action).
  3. **Undo** and **Redo** are the commands to perform an undo/redo action. Note that there are two buttons 'Undo' and 'Redo' in the main toolbar as well that perform the same actions and will be introduced later.
- **Help** menu
  This menu as shown in Figure 10.5 provides help information about BESO2D. The **Getting Started** command gives a 'Getting Started' document that guides the user step by step

Cut	Ctrl+X
Copy	Ctrl+C
Paste	Ctrl+V
Undo	Ctrl+Z
Redo	Ctrl+Y

Figure 10.4   The Edit menu.

Figure 10.5   The Help menu.

Figure 10.6   The toolbar list.

through simple examples. The **About** command shows a message box which contains brief information about this program and contact details for future upgrades.

10.3.3   Toolbar Area

The Toolbar area (part B in Figure 10.2) contains three sections: **Main Toolbar**, **Draw/Run Toolbar** and **View Toolbar**. The user can click on any place in the toolbar area or the menu bar using the right button of the mouse to pop up a list shown in Figure 10.6. These three toolbars can be switched on and off by clicking the corresponding item in the list. After starting the program, the Main Toolbar and Draw/Run Toolbar will be initially switched on whereas the View Toolbar will be automatically switched on after an FEA or BESO optimization run is finished. When the user moves the mouse onto any button in the Toolbar area, a tool tip will pop up to provide an explanation of the function of the button.

- **Main Toolbar**

  The Main Toolbar is shown in Figure 10.7 and contains 10 buttons as explained below.

  1. The **Zoom in** button allows the user to enlarge the relative size of the model compared to the current display area.
  2. The **Zoom out** button allows the user to reduce the relative size of the model compared to the current display area.
  3. The **Zoom factor** button calls a pop-up dialogue shown in Figure 10.8 where the user is able to specify the zoom factor for the current display area.
  4. The **Save the display area to an image file** button allows the user to save the display area to a '.png' or a '.bmp' file.

Figure 10.7   The Main toolbar.

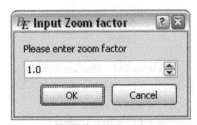

Figure 10.8   Dialogue for specifying the zoom factor.

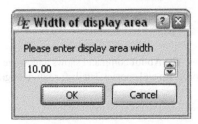

Figure 10.9   Dialogue for specifying the display area width.

5. 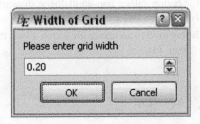 The **Specify display width** button allows the user to specify the width of the display area according to the absolute size of the model that the user intends to create. The dialogue in Figure 10.9 will pop up when this button is clicked on.

6. The **Show grid** button is used to switch on or off a grid in the display area. When the button is in the 'pressed-down' status, the grid is switched on and the mouse will be snapped to the nearest grid points. The grid may assist the user in locating the mouse precisely while constructing a model.

7. By clicking on **Grid width** button, the dialogue shown in Figure 10.10 will pop up where the user can specify the width of the grid in the display area (i.e. the distance between two adjacent grid points).

8. The **Clear screen** button is used when the user wishes to remove all items in the display area.

Figure 10.10   Dialogue for specifying the grid width.

Figure 10.11   The Draw/Run toolbar.

9. The **Undo** and **Redo** buttons do not have icons. The displayed texts on these two buttons will change according to the action to be un-done or re-done, e.g. the text 'Undo Added a rectangle' will appear on the Undo button after a rectangle is drawn.

- **Draw/Run Toolbar**
  This toolbar, shown in Figure 10.11, contains the buttons for drawing the model, applying loading and boundary conditions and performing FEA and topology optimization. The 19 buttons on this toolbar are explained below.
  1. ✓ This is the **Draw designable region** button. When this button is pressed, all the geometries drawn afterwards will be recognized as designable for BESO optimization, i.e. material in these regions will be removable.
  2. ⊘ When the **Draw non-designable region** button is pressed, geometries drawn afterwards will become nondesignable, i.e. the material in these regions will always remain in the structure.
  3. ━ When the **Subtraction** button is pressed, the user can create holes in previously drawn geometries.
  4. ⬏ This is the **Select region** button. When this button is pressed, the program becomes ready for the user to select an item, e.g. an element, a point load or a constraint for further action such as copy or delete. After being selected, a constraint or force can be even moved by mouse-dragging or be removed by pressing 'Delete' key.
  5. ▢ The **Rectangle** button enables the user to draw a rectangle in the display area by clicking the mouse at one corner of the rectangle, dragging it to another corner and releasing it.
  6. ⬯ The **Ellipse** button can be used to create an ellipse in the display area. The usage of this button is similar to that of the **Rectangle** button.
  7. ◯ The **Circle** button enables the user to draw a circle in the display area by pointing at the centre and a point on the perimeter through mouse clicks.
  8. ⌂ The **Polygon** button can be used to create geometry of any shape. The polygon vertices can be drawn one-by-one using the mouse. To close the polygon, one needs to press the 'Ctrl' key and left-click the mouse to input the final vertex.
  9. ▦ This is the **Generate finite element mesh** button. By clicking on this button, the 'Generate mesh' dialogue shown in Figure 10.12 will appear, where the user can input the distance between two adjacent nodes to define the size of the mesh. Clicking on the 'Generate' button will dismiss the dialogue and the program will start to generate a mesh.
  10. ▣ The **Delete elements from mesh** button can be used to delete individual elements from the meshed structural model. Simply click on this button and then click on any element that needs to be removed.
  11. ╫ The **Add new elements to mesh** button can be used to add elements. By clicking on this button and then click on any empty place, an element will be added there. Note that this function is only valid after a finite element mesh has been generated.

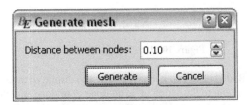

Figure 10.12   Dialogue for generating a finite element mesh.

12. ▥ The **Delete entire finite element mesh** button is used to remove the finite element mesh.
13. ▨ The **Apply constraint to individual nodes** button is used to add a constraint to the degree of freedom on individual nodes. Simply click this button to call the 'Constraint' dialogue shown in Figure 10.13, select the degree of freedom to be fixed and then click 'OK' to dismiss this dialogue; then in the display area, click on the nodes where this constraint is to be applied.
14. ▨ The **Apply constraint to multiple nodes** button is similar to last button, but with the present button the user can drag the mouse to include multiple nodes to which the same constraint is applied.
15. ⟋ The **Apply point load** button is used to apply a concentrated load. By clicking on this button, the 'Point load' dialogue shown in Figure 10.14 will appear where the user can define the force components. Click on 'OK' to dismiss the dialogue and apply this force to the nodes using the mouse.
16. ▥ The **Apply distributed load** button is similar to the last button, but with the present button the user can drag the mouse to include multiple nodes to which the same load is applied.
17. ⟍ The **Set material property** button enables the user to specify the material properties for the structural model. By clicking on this button, a dialogue shown in Figure 10.15 will appear where the user can input Young's modulus, Poisson's ratio and the mass density.

Figure 10.13   Dialogue for applying a constraint.

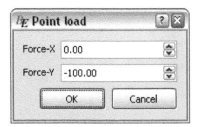

Figure 10.14   Dialogue for specifying point load.

18. $^{F_{EA}}$ When the **Carry out FEA** button is clicked on, the program starts to perform a finite element analysis on the meshed structural model. If the model is not yet meshed, clicking on this button will trigger no action.

19. ^{B_E} By clicking on the **Start BESO** button, a 'Start BESO' dialogue shown in Figure 10.16 will appear where the user can specify BESO parameters. Clicking on the 'OK' button in the dialogue will start the BESO optimization process. The computation may take a while depending on the size of the structural model.

- **View Toolbar**

  The View Toolbar shown in Figure 10.17 is switched off initially and will be automatically switched on after an FEA or optimization run. This toolbar can also be manually switched on any time while running BESO2D. There are six buttons in this toolbar designed for viewing the results of FEA and BESO optimization.

  1. ▬▬ The **Show displacements** button allows the user to view the deformed structure after being analysed. Press down the button to switch on the deformation and release the button to view the un-deformed structure. The deformation shown is automatically scaled to an appropriate level so that it can be seen clearly regardless of the actual magnitude of the deformation.

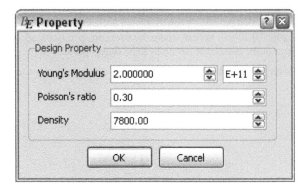

Figure 10.15   Dialogue for specifying material properties.

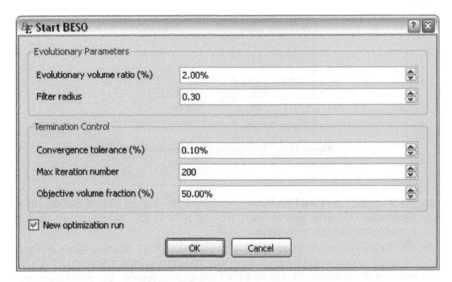

Figure 10.16 Dialogue for specifying BESO parameters.

2. 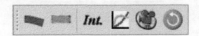 When the **Show stresses** button is pressed down, the von Mises stress distribution will be displayed on the structure and the von Mises stress levels will be given on the right hand side of the display area as shown in the Figure 10.18.

3. *Int.* The **Show intermediate design** button is used to display any intermediate design after an optimization run is finished. By clicking on this button, the 'Intermediate design' dialogue shown in Figure 10.19 will appear where the user can select an iteration number and then click on 'Show' to display the design of the specified iteration.

4. The **Plot BESO histories** button can be used to call a window showing the evolution histories of the volume fraction and the mean compliance.

5. The **Movie** button allows the user to sequentially view all the intermediate designs of the BESO optimization process. When the button is clicked on, the 'Run movie' dialogue shown in Figure 10.20 will appear where the user can specify the speed of playing the movie in terms of the number of iterations to be shown per second.

6. The **Revert** button allows the user to revert the optimized structure to its initial design.

Figure 10.17 The View toolbar.

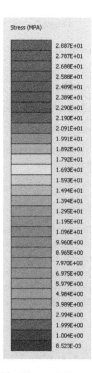

Figure 10.18   The von Mises stress levels.

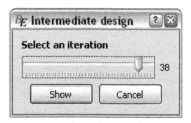

Figure 10.19   Dialogue for selecting an intermediate design.

Figure 10.20   Dialogue for specifying the speed of playing the movie of intermediate designs.

Zoom x 1   (X, Y) = (6.94, 2.55)   Element ID 553

Figure 10.21   The status bar.

10.3.4   Display Area and Status Bar

The Display area (part C in Figure 10.2) is where the user draws structures and applies loading and boundary conditions. The dimensions of the display area can be changed by using the [A₪] button. FEA and BESO results are also shown in this area. When the ⋮⋮⋮ button is switched on, the display area will be filled with grid points.

10.3.5   Status Bar

The Status bar (part D in Figure 10.2) contains three message blocks shown in Figure 10.21. The first block shows the current zoom factor, the second block indicates the current position of the mouse in the display area and the third block identifies the element at which the mouse is currently pointing. The second message block is helpful when the user wish to precisely position the mouse and to apply a constraint/load to a desired location. Note that BESO2D has not specified any particular units for the length, force etc. The users may choose appropriate and consistent units themselves.

10.4   Running BESO2D from Graphic User Interface

In this section we shall demonstrate, step-by-step, how to use the BESO2D GUI window, from constructing the structural model to obtaining the final optimized topology. The procedure includes the following steps:

1. Drawing the design domain of a structure.
2. Generating a finite element mesh of the design domain.
3. Specifying boundary conditions, loading conditions and material properties.
4. Performing FEA on the meshed model and showing the analysis result.
5. Performing BESO optimization.
6. Viewing the final optimal design and the evolution histories.
7. Continuing optimization from an obtained design.

To illustrate these steps, a typical Michell type structure shown in Figure 10.22 is used as an example.

10.4.1   Drawing the Design Domain of a Structure

This step comprises the following actions:

• Start BESO2D.exe.

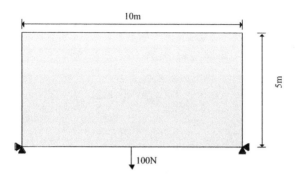

Figure 10.22   Design domain of a Michell type structure.

- Set the display area width: click on the ⊞ button in the **Draw/Run Toolbar** and enter 15.0 in the 'Width of display area' dialogue and click on 'OK'.
- Draw a rectangle: click on the □ button in the **Draw/Run Toolbar**, locate the mouse at the position of (1.0, 2.0) (see the current position of the mouse in the Status bar), press the left mouse button, drag the mouse to the position of (11.0, 7.0) and release the mouse. A rectangle is obtained in the display area as shown in Figure 10.23.

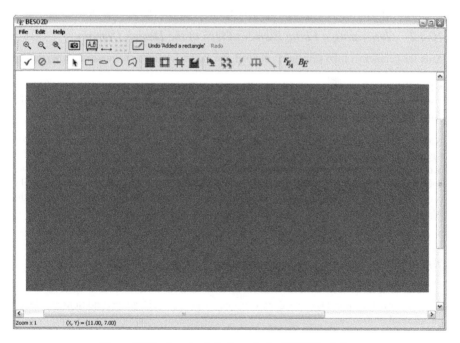

Figure 10.23   A rectangle is drawn in the BESO2D window.

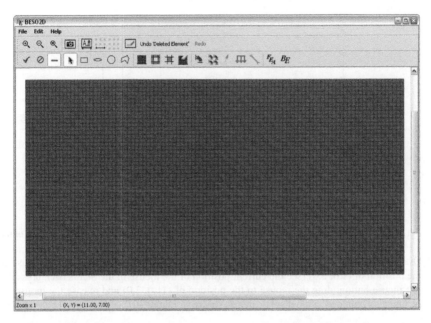

Figure 10.24   A finite element mesh is generated for the structural model.

10.4.2   Generating a Finite Element Mesh of the Design Domain

After the geometry of the structure is defined, the model is ready to be meshed in this step.

- By clicking on the ▦ button in the **Draw/Run Toolbar,** the 'Generate mesh' dialogue will appear. Input 0.10 for 'Distance between nodes' and then click on 'Generate'. Now the meshed structural model is obtained as shown in Figure 10.24.

Note that when the mesh is generated, the corners of the structure may sometimes be shifted slightly to nearby locations and thus the geometry of the design domain is changed. This occurs when the length of the structure divided by the distance between two adjacent nodes does not yield an integer. Therefore, it is recommended to check of the coordinates of the corner nodes after mesh generation. If necessary, the user may add or delete elements at the edges to maintain the intended dimensions for the structure. Deleting or adding elements can be done by clicking on ▦ or ╫ buttons and then click the mouse at the relevant location.

10.4.3   Specifying Boundary Conditions, Loading Conditions and Material Properties

After the structural model is meshed, we can apply boundary and loading conditions and define the material properties.

- By clicking on the ⮝ button in the **Draw/Run Toolbar**, the 'Constraint' dialogue will appear. Select 'All degrees of freedom fixed' and click on 'OK'. Now the dialogue will

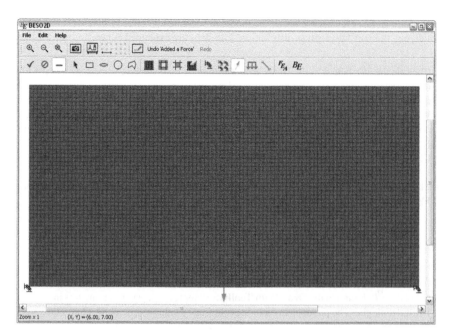

Figure 10.25   Structural model with boundary and loading conditions.

disappear. Move the mouse close to the two bottom corners of the meshed rectangle and click the left mouse button to add the constraint to the two corner nodes. Note that one does not need to position the mouse exactly on the node but can click on a position which is near enough – the constraint will be automatically applied to the nearest node.

- By clicking on the *⁄* button in the **Draw/Run Toolbar**, the 'Point load' dialogue will appear. Input the force components as 0 for 'Force-X' and -100.0 for 'Force-Y'. Click on 'OK' and the dialogue will disappear. Move the mouse to the position of (6.0, 7.0) (the mouse position can be seen from the Status bar) and press the left mouse button to apply the force. Note that the force will be applied to the node nearest to the position of the mouse. The structural model with boundary and loading conditions is shown in Figure 10.25.

- By clicking on the ⤣ button in the **Draw/Run Toolbar**, the 'Property' dialogue will appear. Enter the material properties such as 2.0E+11 for 'Young's modulus', 0.3 for 'Poisson's ratio' and 7800.0 for 'Density'. Note that in current BESO2D the stiffness optimization is carried out without considering the gravity loading. Therefore, the density (i.e. mass density) does not affect the optimal design.

10.4.4   Performing FEA on the Meshed Model and Showing the Analysis Result

Now the model is ready for finite element analysis and BESO optimization. Although an FEA before optimization is not compulsory, it is always helpful and therefore strongly recommended to perform an FEA to check whether the model is constructed correctly by checking the FEA result, e.g. the von Mises stress distribution. The steps are explained below.

- Click on the F_{E4} button in the **Draw/Run Toolbar**. A message box to indicate that the FEA is running pops up. The computation time depends on the total number of elements and the computer hardware configurations.
- When the FEA is finished, the von Mises stress distribution will be automatically displayed on structural model. The stress levels can be found on the right hand side of the display area. Note that the **View Toolbar** now appears in the toolbar area.
- Click on ▬ and ▦ buttons in the **View Toolbar** to switch on or off the deformation and stress distribution.
- The initial model for the Michell type structure has been correctly constructed if the obtained FEA result is similar to that shown in Figure 10.26.

10.4.5   Performing BESO Optimization

To perform BESO on the structure, follow these steps.

- Click the B_E button in the **Draw/Run Toolbar.** Enter (or accept) the parameters in the 'Start BESO' dialogue. The parameters include evolutionary volume ratio (ER), filter radius (r_{min}), convergence tolerance τ, maximum iteration number and the objective volume fraction.
- The evolutionary volume ratio controls the volume fraction change between two consecutive iterations. In this example, set $ER = 2.0\ \%$.

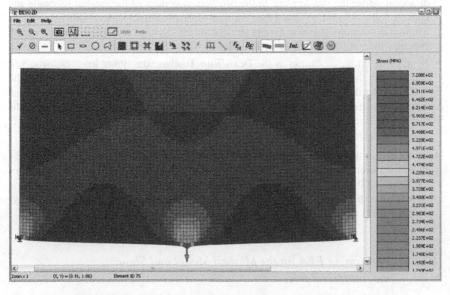

Figure 10.26   Displacement and stress distribution of the initial model.

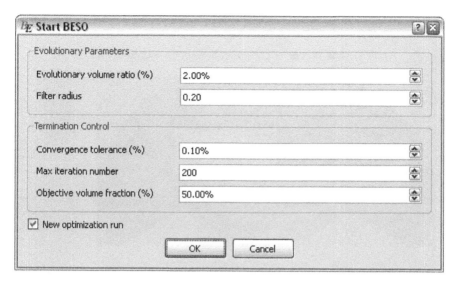

Figure 10.27   The BESO parameters used in this example.

- To overcome checkerboard pattern and mesh-dependency problems, the filter radius is recommended to be 2~3 times the element size. In this example, set the filter radius to 0.20.
- The convergence tolerance gives an allowable convergence tolerance such as 0.10 % in this example.
- The maximum iteration number is set to be 200 in this example.
- The objective volume fraction defines the final material usage as a percentage of the material in the design domain. Set 50.0 % for this example.
- Make sure the 'New optimization run' checkbox is checked. Now the 'Start BESO' dialogue shown in Figure 10.27 will appear.
- Now click on the 'OK' button. The optimization starts immediately and a 'BESO running' message pops up.
- After the pre-processing, the 'BESO running' message box will display information indicating the current iteration number. During the optimization process, the intermediate designs are shown in the display area. Note that if the ▬▬ button has been pressed down, the real-time stress distribution will be displayed on the structure like Figure 10.28. The user can see how the structure is evolving on the screen.

10.4.6   Viewing the Final Optimal Design and the Evolution Histories

- After a short while, a 'BESO finished' message box will show up indicating the end of the optimization run with the message 'Optimization converged to a solution'. Click on 'OK' to dismiss the message and the final optimized design is shown in the display area as in Figure 10.29.

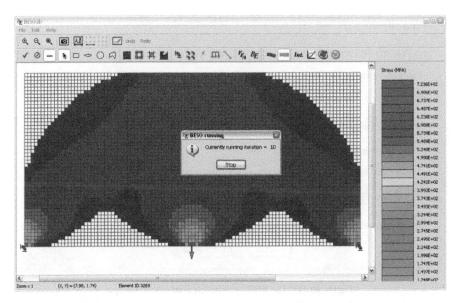

Figure 10.28   An intermediate design during the optimization process.

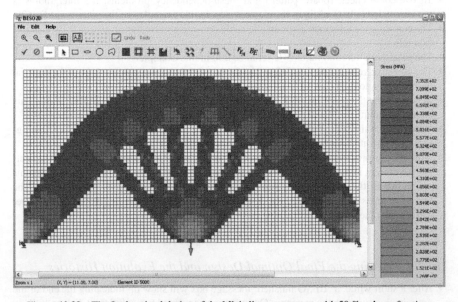

Figure 10.29   The final optimal design of the Michell type structure with 50 % volume fraction.

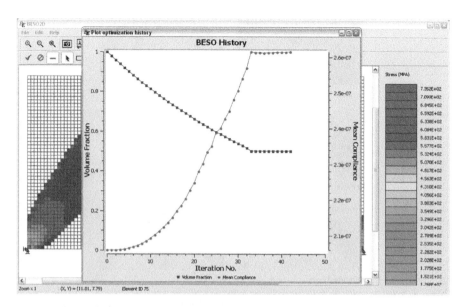

Figure 10.30   Evolution histories of the volume fraction and the mean compliance.

- You are now ready to review any intermediate designs and the evolution histories. Click on the ⬚ button to view the histories of the volume fraction and the mean compliance as shown in Figure 10.30.
- By clicking on the *Int.* button, the 'Intermediate design' dialogue will appear. Select a desired iteration number and click 'Show', the corresponding intermediate design will be displayed. The intermediate designs can be saved to picture files in '.png' or '.bmp' format using the ⬚ button in the toolbar area.
- Click on the ⬚ button and select a frame display speed. A speed of 1 iteration/second is recommended for clearly showing the whole optimization process. Then click on 'Run' to start the movie. Some movie sample frames are shown in Figure 10.31.

10.4.7   Continuing Optimization from an Obtained Design

Sometimes the optimization run stops before the solution converges due to reasons such as that the maximum iteration number has been set too small. BESO2D may continue the optimization process from the design of a previous optimization run.

In this example, we demonstrate how to further optimize the Michell type structure from the previous solution and obtain a final optimal design with 15.0 % volume fraction.

- Click on the *BE* button in the 'Draw/Run Toolbar'. In the 'Start BESO' dialogue, input 15.0 % for the objective volume fraction.
- Uncheck the 'New optimization run' checkbox and click 'OK'.

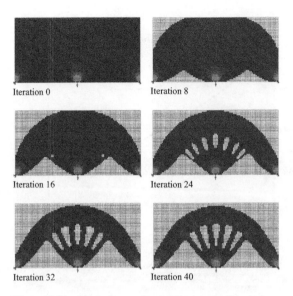

Iteration 0   Iteration 8

Iteration 16   Iteration 24

Iteration 32   Iteration 40

Figure 10.31   Movie frames showing the intermediate designs.

- After a short while, the optimization run will stop and the final design with 15.0 % volume fraction will be obtained as shown in Figure 10.32.
- Follow the same steps in 10.4.6 to view the intermediate designs and evolution histories.

10.5   The Command Line Usage of BESO2D

The command line usage of BESO2D is suitable for higher level users. The BESO2D engine, i.e. the core executable file 'BESO.exe' which performs FEA and optimization can be used as a standalone program independent from the GUI and may be run in the command line. The advantage is that that running the BESO2D engine from the command line is faster than running the program through the BESO2D GUI since the engine only executes the necessary actions (mainly number crunching). The command line usage is especially useful when a large model is dealt with. More importantly, it allows the user to import an initial design created by a third-party program which might be more powerful in generating meshes than the BESO2D GUI. The format of the model file accepted by the BESO2D engine is simple and easy to understand. After an optimization run is complete, the user can view the final optimal design in the BESO2D GUI window by importing the design model file into the program and view the evolution history data stored in a text file 'Result.txt'.

10.5.1   Calling the BESO2D Engine

An example of calling the BESO2D engine is as follows:

D:\BESO2D\BESO Model.txt Parameters.txt

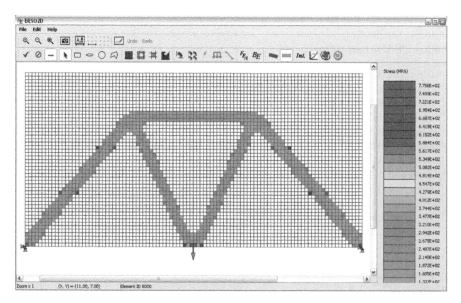

Figure 10.32   The final optimal design of the Michell type structure with 15.0 % volume fraction.

The BESO2D engine takes two arguments to get started. They are the names of two text files: the first is the model file of the initial structure and the second is the file that contains the BESO parameters.

10.5.2   The Model File Format Accepted by the BESO2D Engine

In order to explain the format of the model file, a typical sample file is introduced below for a coarsely meshed model shown in Figure 10.33.

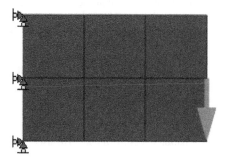

Figure 10.33   A coarsely meshed model for explaining the model file format.

- - - - -Model file 'Sample.txt' begins here- - - - -
*model<Sample, 2, 1 >
*title< QUAD4 User defined model >

*node< 1, 0, 3.5, 2.0, 0 >
*node< 2, 0, 4.0, 2.0, 0 >
*node< 3, 0, 4.5, 2.0, 0 >
*node< 4, 0, 5.0, 2.0, 0 >
*node< 5, 0, 3.5, 2.5, 0 >
*node< 6, 0, 4.0, 2.5, 0 >
*node< 7, 0, 4.5, 2.5, 0 >
*node< 8, 0, 5.0, 2.5, 0 >
*node< 9, 0, 3.5, 3.0, 0 >
*node< 10, 0, 4.0, 3.0, 0 >
*node< 11, 0, 4.5, 3.0, 0 >
*node< 12, 0, 5.0, 3.0, 0 >

*quad4< 1, 2, 2, 1, 2, 6, 5, 0 >
*quad4< 2, 2, 2, 2, 3, 7, 6, 0 >
*quad4< 3, 2, 2, 3, 4, 8, 7, 0 >
*quad4< 4, 2, 2, 5, 6, 10, 9, 0 >
*quad4< 5, 2, 2, 6, 7, 11, 10, 0 >
*quad4< 6, 2, 2, 7, 8, 12, 11, 0 >

*force< 1, 8, 0, 0, -10 >
*constraint< 1, 1, 0.0, 0.0 >
*constraint< 2, 9, 0.0, 0.0 >
*constraint< 3, 5, 0.0, 0.0 >

*property<1, 200.0, 0.3, 7.8 >
*property <2, 200000000000.0, 0.3, 7800 >
*thickness<1.0 >
- - - - -Model file 'Sample.txt' ends here- - - - -

In this model file, lines beginning with '*' are considered valid. Different words following '*' form different tags. The contents within '< >' are the parameters which are separated by ',' and needed for specific tags. The following eight tags are used for creating a complete model file.

- *model<model name, 1st reserved parameter ('2'), 2nd reserved parameter ('1')>
  This tag provides the basic information of the model. The first parameter is the model name. The second and the third parameters are reserved and fixed as 2 and 1 as used in the above sample file.
- *title<title>
  This tag gives further information about the model. The user may choose to input any text to describe the model.
- *node<node ID, coordinate system ('0'), X, Y, reserved parameter ('0')>

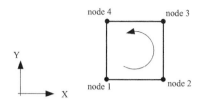

Figure 10.34   The order of nodes in each element.

This tag defines a node. The first parameter is the node ID, the second parameter always takes the value of 0 indicating that the global coordinate system is used, the third and fourth parameters are the X and Y coordinates of this node. The last parameter is reserved as '0' and can be absent. Note that each node ID may only be used once and the IDs should be numbered with an increment of 1 for the whole model.

• *quad4< element ID, PID, reserved parameter ('2'), node 1, node 2, node 3, node 4, designability >
This tag defines an element. The first parameter is the element ID. Similar to the node IDs, the element IDs must be defined sequentially with an increment of 1. The second parameter is the property ID, designating the material used (see the description for '*property' below). The third parameter is reserved and always takes the value of 2 as shown in the sample file. The next four parameters are the IDs of the four nodes of this element. The last parameter determines the design ability of the element, namely 0 for being designable and 1 for being nondesignable. Note that the IDs of the four nodes in each element should be given in a counter-clockwise order, as shown in Figure 10.34.

• *force< force ID, node, coordinate system ('0'), X component, Y component >
This tag defines a force. The first parameter is the force ID, the second parameter is the ID of the node where the force is applied. The third parameter always takes the value of 0 indicating the global coordinate system. The last two parameters are the X and Y components of the force.

• *constraint< constraint ID, node, X freedom, Y freedom >
This tag defines a constraint. The first parameter is the constraint ID, the second parameter is the ID of the node where the constraint is applied. The last two parameters are the specified displacements in the X and Y directions. In two extreme cases where the specified displacement is '0.0' and 'free', it defines a fixed degree of freedom and a free degree of freedom, respectively.

• *property<PID, E, u, rho >
This tag defines a material. The first parameter is the material ID (also known as property ID). The next three parameters are Young's modulus, Poisson's ratio and the mass density, respectively.

• *thickness< thickness >
As plane stress conditions are assumed in BESO2D, this tag defines the thickness of the structure. Unit thickness will be assumed if this tag is absent.

Note that such a model file can also be exported by the BESO2D GUI. Simply create a model in BESO2D GUI and then select the menu item 'File → Export Model' to save the created model to a model file.

10.5.3   Format of BESO Parameter File

A typical parameter file is given below.

```
- - - - -Parameter file 'Parameters.txt' begins here- - - - -
*Model<Test>
*EvoVolRatio<0.02>
*FilterRadius<0.2>
*ConvTolerance<0.001>
*MaxIter<200>
*ObjVolFraction <0.5>
- - - - -Parameter file 'Parameters.txt' ends here- - - - -
```

- *Model< model name >
  This tag specifies the model name of the structure to be optimized. The intermediate designs will be stored in files whose names are composed of the model name and a three-digit iteration number, e.g. the intermediate design of iteration 38 will be stored in a file named Test038.txt, if the model name of 'Test' is specified.
- *EvoVolRatio< ER >
  This tag defines the evolutionary volume ratio. A typical value is 0.02 (i.e. 2.0 %).
- *FilterRadius< r_{min} >
  This tag defines the filter radius. A typical value is two to three times of an element size.
- *ConvTolerance< τ >
  This tag defines the allowable tolerance error for convergence. A typical value is 0.001 (i.e. 0.10 %).
- *MaxIter< maximum iteration number >
  This tag defines the maximum number of iterations.
- *ObjVolFraction < volume fraction >
  This tag defines the objective volume fraction of the final optimal design. Its value is between 0 and 1, e.g. 0.5 for 50 % volume fraction.

10.5.4   Result File of an Optimization Run

After the optimization run is finished, a text file named 'Result.txt' can be found in the folder where the BESO2D engine is. The sample result file is given below.

```
- - - - -Result file 'Result.txt' begins here- - - - -
Bi-directional Evolutionary Structural Optimization: BESO2D Version 1.0
Model: Test    Objective: Maximizing stiffness
Starting Date: 09/13/09 Time: 17:18:07
```

Iter	VolFrac	meanCompliance	Time
Start			17:18:07
0	1.000000	2.058341E-007	17:18:09
1	0.979620	2.058365E-007	17:18:10

2	0.959640	2.058699E-007	17:18:12
3	0.939660	2.059636E-007	17:18:14
4	0.920080	2.061125E-007	17:18:15
5	0.900899	2.063617E-007	17:18:17

· · · · · ·

· · · · · ·

- - - - -Result file 'Result.txt' ends here- - - - -

Here the results of the first six iterations are presented. The first column of the data identifies the iteration number. The second and third columns show the volume fraction and the mean compliance, respectively. The last column records the time when an iteration is finished.

10.6   Running BESO2D from the Command Line

In this section, a step-by-step guidance for running the BESO2D from a command line is presented. This includes optimizing a structure from an initial design and continuing optimization from a previously obtained design solution.

10.6.1   Optimize a Structure from an Initial Design

Before optimizing a structure using the BESO2D engine, we need to have a working folder where we have the authority for file operations such as reading, writing and deleting. Various files will be produced by BESO.exe including the model files of the intermediate designs, a 'Result.txt' recording the evolution histories and a brief report in 'Report.txt'. In this section, it is assumed that the working folder is 'D:\BESO2D\'.

- Edit the model file of the initial structure and the parameter file (manually or using a program such as BESO2D GUI or a third-party software package). In this section, it is assumed that we have the model file named 'Test.txt' for the previous initial Michell type structure and the sample parameter file 'Parameters.txt' in 10.5.3. Now we can start optimizing the structure by calling BESO2D in a command line.
- Empty the working folder 'D:\BESO2D\'. Put the BESO2D engine 'BESO.exe', 'Test.txt' and 'Parameters.txt' into the working folder.
- Open the command line window: click on 'Run...' in the Windows Start menu, input 'command' in the edit field and then click on 'OK' button.
- Change the current directory to the folder where 'BESO.exe', 'Test.txt' and 'Parameters.txt' have just been stored by typing 'D:', then 'cd D:\BESO2D\' in the command line window as shown in Figure 10.35.
- Under the prompt 'D:\BESO2D>', type 'BESO Test.txt Parameters.txt' and press 'Enter' key. Now the program starts running.
- Some information will be displayed in the command line window, e.g. the iterations completed. When the program stops due to the maximum iteration number being reached or the solution having converged, a message 'END OF BESO' will be displayed in the command line window.

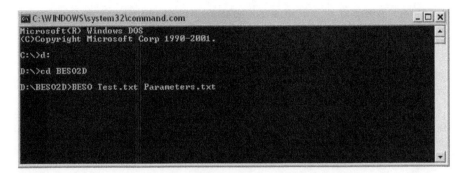

Figure 10.35   Running BESO2D engine from a command line.

- Now the final design can be imported to BESO2D GUI to be viewed. Under the 'File' menu of the BESO2D GUI window, click 'Import Model', go to the working folder 'D:\BESO2D' and select the model file with the largest iteration number which contains the last design. Click on 'OK' to open this design. Then the final design will be displayed in the BESO2D GUI window as shown in Figure 10.36. Note that the designs in the last few iterations are similar due to the convergence of the solution.

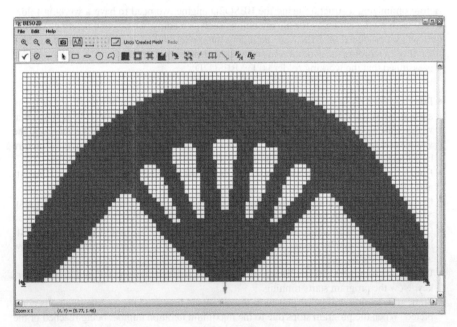

Figure 10.36   Displaying the solution with 50.0 % volume fraction obtained from a command line optimization.

- View the evolution histories of the volume fraction and the mean compliance of the structure by opening the 'Result.txt' in the working folder.
- A brief report showing the total number of iterations used and the procedure termination criterion can be found in the text file 'Report.txt' in the working folder.

10.6.2   Continuing Optimization from a Previously Obtained Design Solution

After an optimization run is finished and an optimal design is obtained, the structure can be further optimized using different optimization parameters. In this example, we change the objective volume fraction from 50.0 % to 15.0 % and further optimize the Michell type structure.

- Open the 'Parameters.txt' in the previous working folder 'D:\BESO2D\', set a new objective volume fraction by changing the corresponding line to *ObjVolFraction <0.15>. Save and close this file.
- Under the command prompt 'D:\BESO2D>', type the same command as before: 'BESO Test.txt Parameters.txt'. Press enter to start the optimization further. Now the BESO2D engine will automatically continue the optimization process from the end design of last run. Again the message 'END OF BESO' will appear once the optimization is finished.
- Open the BESO2D GUI window, import the model file of the final design to view the final optimal design as shown in Figure 10.37.

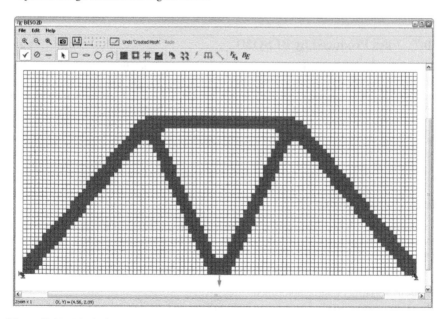

Figure 10.37   Displaying the solution from command line optimization for the new objective volume fraction of 15.0 %.

Table 10.1 Files produced by BESO2D.

File name	File description
<Model>NNNalp.bin	Binary file storing the elemental sensitivity numbers of iteration NNN.
Filter.bin	Binary file storing the filter scheme.
Result.bin	Result file that stores the histories of iteration number, volume fraction and mean compliance, in binary format.
<Model>NNN.evm	Binary file storing the element mean von Mises stresses of the corresponding intermediate design. Used for plotting the stress distribution.
<Model>NNN.ndp	Binary file storing the nodal displacements of the corresponding intermediate design. Used for plotting the structural deformation.
<Model>DOMAIN.txt	Formatted model file of the initial design.
<Model>NNN.txt	Model file of intermediate design of iteration NNN.
Iteration.txt	Text file recording the iteration number that has completed.
Report.txt	A brief report for an optimization run.
Result.txt	Result file storing the histories of iteration number, volume fraction and mean compliance, in ASCII format.

- View the evolution histories by opening the 'Result.txt' in the working folder 'D:\BESO2D'. The additional result data from the continued optimization run are appended to the previous results.
- A new report is found in 'Report.txt'.

10.7   Files Produced by BESO2D

This section describes files that are produced during optimization by BESO2D. Some of the files include the model name in the main part of the file names. For a clear explanation of the file names, we now assume the model name to be <Model> and the iteration number to be NNN. The produced files together with a description of each file are listed in Table 10.1.

Note that the file names are changed consistently with the model name, e.g. if we change the model name tag to *Model<Sample> in the parameter file, the model file of the intermediate design in iteration 23 becomes Sample023.txt.

10.8   Error Messages

Some errors may occur while running the program. These may be caused by inappropriately constructed structural model or other reasons. For a common error, an error message will appear.

- 'Stiffness Matrix Factorization Failed'
  This error message may appear during finite element analysis. The error is most likely caused by an inappropriately constructed structural model, e.g. with insufficient supports applied to the structure.

- 'Error building filter scheme, filter radius is too small'
  When the filter radius is smaller than half the size of one element, this error message appears. In this case, the filter radius needs to be increased so that at least one element is covered by the circular domain centred at the element centroid with a radius being equal to the filter radius.
- 'Error reading element sensitivities'
  This error may occur only when an optimization run is continued from a previously obtained design from the command line. The error is caused by absence of the <Model>NNNalp.bin file (see the last section for file description). Therefore when running the BESO2D engine from the command line, one needs to be careful not to delete any file in the working folder unless he/she intends to start a new optimization run.
- 'Input file does not exist. Building model failed'
  This error is most likely to occur when continuing an optimization run using the BESO2D engine from the command line. This is caused by the removal of the model file of the final design of last optimization run.
- 'Error reading following BESO parameters'
  This error may occur when running BESO2D from the command line. Note that all six tags are needed to start an optimization run: *Model, *EvoVolRatio, *FilterRadius, *ConvTolerance, *MaxIter and *ObjVolFraction. Furthermore, the specified parameters must be meaningful. For example, an objective volume fraction outside the range between 0 and 1 will cause this error.

If the reader has any difficulties in obtaining or using BESO2D, or has queries about updates of the program, please contact us by email at the following addresses:

Zhihao.Zuo@rmit.edu.au   (Z.H. Zuo)
Huang.Xiaodong@rmit.edu.au   (X. Huang)
Mike.Xie@rmit.edu.au   (Y.M. Xie)

Author Index

Evolutionary Topology Optimization of Continuum Structures: Methods and Applications   Xiaodong Huang and Mike Xie
© 2010 John Wiley & Sons, Ltd

Subject Index

Printed and bound by CPI Group (UK) Ltd, Croydon, CR0 4YY

16/04/2025

14658442-0010